Softwood Drying

Enhancing Kiln Operations

• Larry Culpepper •

Miller Freeman Books
San Francisco

Published by Miller Freeman Books
600 Harrison Street, San Francisco, CA 94107

Miller Freeman Books

Cover Design: Brad Greene
Interior Design: Brad Greene

ISBN 0-87930-581-9

Printed in the United States of America
00 01 02 03 04 5 4 3 2 1

To my kids,

Daniel and Claire.

Thanks, guys.

You make it all worthwhile.

❖

Table of Contents

Introduction

"People don't choose their careers; they are engulfed by them."

—John Dos Passos

It is hard to believe that ten years have passed since publication of "High Temperature Drying—Enhancing Drying Operations". What a trip this has been. The reception of that book was very gratifying and the many kind comments of readers are the inspiration for this—what began as a second edition to the first.

"High Temperature Drying" was, an continues to be, very successful. However, the title branded me (particularly in certain circles) as having a limited range of experience. Nothing could be further from the truth and I trust that this text dispels the myth.

This book does build on the first, adding another ten years of kilns and experience to the mix. During its development I have sought to blend in experiences from the many different regions of the country, and out of it, that my travels have taken me. The northeast, mid west, Great lakes, northwest, the south (of course!), Canada, Mexico, and Russia all have had an influence. It is interesting that, aside from the obvious species differences, kiln drying is pretty much the same wherever I go. They all share similar techniques. They all make the same mistakes. Through it all, I have endeavored to explain not only the how, but the why. I firmly believe that if you understand the fundamentals of what you are doing then running the basic dry kiln is easy.

I teach several drying classes a year and always begin each by explaining what I consider to be the four fundamental elements of kiln drying. It is my conviction that if you are able to grasp the deeper meaning of what seem like simple concepts, understand and dissect them, and then apply them to your daily routine as well as each problem you encounter you have the potential to be an outstanding kiln operator. Not just good—outstanding. As you read through this book you will find that these fundamentals are woven throughout.

Rather than have to stop and explain their application each time they pop up it seems prudent to list them here

The first fundamental is that, reduced to its minimum, kiln drying is the application of Heat to evaporate the water in wood. The total amount of heat required to evaporate this water is dependent on many things; the wood species and its beginning moisture content, the type of kiln, the heating medium, building design, and so on. All the airflow in the world will not overcome a lack of heat and when the heat stops, so does drying. Keep in mind that each individual board foot of lumber in your kiln requires a finite amount of heat to dry down to the desired moisture content.

The second fundamental helps us understand how to deliver that heat and is summed up in the following simple formula

$$\text{HEAT TRANSFER} = f\ (\text{TEMPERATURE} \times \text{AIR VELOCITY})$$

Heat transfer (delivery) is a function of the temperature of the air times its velocity. Now, imagine an entire charge of lumber loaded and ready to enter the kiln. Imagine also that all of the lumber, each board foot, has the same moisture content. We know from the first fundamental that we need only to deliver the same amount of heat to each board foot in order to dry the lumber to the desired moisture content. By this formula, if the air velocity is uniform throughout our load, we need only maintain a constant temperature to achieve this. The rub comes when you finally realize that air velocity can never be uniform throughout the load and here is where the art and understanding merge. This second fundamental helps us understand why there are moisture content variations in our kilns and points to why such concepts as multi-zoning, exiting air heat control, and plenum baffles work.

In short, it helps us understand the reason for the drying variation we experience in running the kilns.

The third fundamental is summed up in a single word—Equilibrium. To be in equilibrium means to be in balance. Everything which happens in kiln drying happens because we upset the equilibrium. Air moves, heat is transferred, moisture migrates all because of the tendency for things (heat, air, moisture) to flow from areas of high concentration to areas of low to regain its equilibrium. The fans build up a higher pressure on one side (static pressure) than on the other and the air moves through the lumber to regain its equilibrium. Turn off the fans and the air flow stops. The boiler firebox is hotter than the pressure vessel where the steam is generated. The coils are hotter than the surrounding air which is in turn hotter than the lumber surface. The breakdown in the flow of heat becomes immediately obvious when the fuel flow stops.

During the active drying cycle the moisture content of the lumber is higher than the surrounding air. In addition, as long as moisture is evaporating from the lumber surface, the core of each board is higher than its surface. During Equalizing and Conditioning steps this relationship is reversed and moisture flows back into the wood. Loose power, and everything grinds to a halt.

Consistency is the fourth fundamental. In all that you do, strive for this ideal. From the log yard to the sawmill and on through to the planing mill look for areas of inconsistency and eliminate them. Cut it, sort it, stack it, load it, dry it and store it all the while maintaining as consistent control as you possibly can. When consistency becomes a habit at your mill 90 percent of your problems will evaporate, allowing you to concentrate on the remaining 10 percent, tweaking and refining as you go along. Do it all day, everyday, and you will discover how easy it is to become an outstanding kiln operator.

Copy these fundamentals down and tack them over your desk as a reminder.

Live them as well as learn them to reap their full benefit.

As before, there have been a variety of people who have helped to influence this book and I would like to recognize them here. To the list of those mentioned in the first book I would like to add the names of Bill and Tracy Smith, Nelson Flippo, Bill Eldridge, Barry Disken, Mark Chenard, Ed Mockler, Brian Faiman, Tom Coxe, Holt Hogan, Preston Warman, Al Raymond, Ron Roberts, and Bill Dawkins

Larry G. Culpepper
August, 1999

Chapter One

Lumber Manufacture and Its Impact on Drying

Be not afraid of going slowly.
Be more afraid of standing still.

Chinese proverb

Why Worry about Lumber Manufacture?

Simple: Poor lumber manufacture costs money. Quality drying is achieved by reducing the variability of factors affecting the drying process. Poor manufacture affects lumber yields, handling costs, energy costs, equipment life, safety, moisture distributions, lumber, and grade recovery by introducing variability. Concern for controlling variability begins at the sawmill with the delivery and unloading of logs onto the yard.

Log Storage

The beginning moisture content of green lumber affects its drying rate.[1] All other factors being equal, boards with high moisture content at the beginning of drying will be higher in moisture content at charge end. Studies have shown that the moisture content of timber from which lumber is produced varies according to many factors.[2] These include:

- species
- tree age
- wood density
- geographic or site location
- height from ground
- heartwood versus sapwood
- storage method

For practical purposes, the only factor we can control is storage method. All logs must be placed into the sawmill on a first-in/first-out basis with every effort made to minimize the storage time variation within a log run. This can be a challenge at northern mills where logs might be stockpiled for months in the bush prior to delivery to the mill. Obviously, this factor may be beyond our control, but we must keep in mind that our ultimate goal is to reduce drying variation in our final products. By working closely with log suppliers, it should be possible to reduce this source of variation.

Drying problems related to storage seem to be greatest at mills that store logs under water sprays. At certain times of the year, I can count on calls from mills who water-store and suddenly cannot control the moisture variation coming out of the kilns. Water-stored logs take longer to dry in a dry kiln. This is probably due to bacterial growth within the cell structure of the log as it is held under storage. As long as we separate water-stored logs from green logs as they are processed through the mill, we can control the problems associated with water storage. However, it is common for the person loading the log deck to unload arriving trucks onto the log deck while feeding water-stored logs. This must be avoided, since it leads to random wet and overdry spots within the charge, which cannot be corrected through the drying schedule. Extensions in drying time will occur with water-stored logs. The lumber from intermixed green logs will be seriously overdry by the time the stored logs are down to an acceptable moisture content.

To improve the effectiveness of your water-sprinkled log yard, make certain your sprinkler system keeps the ends of logs wet. This reduces the incidence of problems from fungal and insect attack. Some species—eastern white pine, ponderosa pine, and to some degree southern pine, as well as others—are susceptible to brown stain in the finished dried product, which is objectionable for many uses. This may be a particular problem with water-stored logs at certain times of the year. Providing aeration to the holding ponds has reportedly been effective in controlling this problem.

Holding logs for long periods of time without rotating them through the system encourages sap decay, fungal growth, and worm holes. The lumber cut from these logs is very difficult to dry. The logs are at varying levels of moisture content and mold will have already penetrated to the heartwood along the rays. Under the right conditions, the lumber cut from this timber will form mold blooms on the surface within hours of manufacture. Lumber that has reached this stage is characterized by deep blue stains and a grayish brown coloration to the wood. You may notice large worm holes on cut surfaces and a rubbery white fungus spreading over the ends and faces of the lumber. This is a sign of advanced sap decay, and it generally excludes the lumber from any of the grades on the basis of strength reduction. Inspect your log yard for the telltale signs: Black streaks on the butt ends of dry

Figure 1.1 Establish and maintain your log storage system on a first-in/first-out basis. Keep the logs you are processing into the mill as evenly aged as possible to avoid problems. Mixing ages is asking for trouble.

stored logs and little piles of fine shavings are an indication that they have been in storage too long. Log yards must be managed on a first-in/first-out basis if at all possible (figure 1.1).

Variations in Lumber Manufacture

Manufacturing all lumber to uniform dimensions eliminates manufacture as a source of drying problems. Unfortunately, lumber size can vary a great deal depending on the set work accuracy of even a single machine. The variation between different machine centers is often extreme.

Check machine centers daily and make corrections as problems occur. Identify substandard equipment and repair or replace it as needed. The long-term solution to this problem is the effective implementation of a quality control (QC) program in the sawmill. A well-run QC program pays back in increased lumber yield through reductions in target size and increased grade recovery. It is well worth the effort for any mill. Guidelines for starting a QC program are beyond the scope of this book. For more information I recommend Terry Brown's text, *Quality Control in Lumber Manufacturing.*[3] In the absence of a quality control program, the kiln operator must be on the lookout for problems in manufacture that affect drying quality and complain, complain, complain until the problems are addressed.

Variations in Thickness

Of all the variations in manufacture, variations in lumber thickness have the greatest impact on drying. These variations affect grade and equipment as well as moisture distributions. Thick lumber takes longer to dry than its thinner neighbor, and variations of 0.125 inch are significant. Measurement of the actual impact of thickness variation is difficult given the many variables that affect final moisture content. However, computer simulations of drying rates indicate that a $\pm$0.10 inch variation in lumber thickness results in a 3.5% MC variation in final moisture content. This variation adds to the many sources of variation common in any drying system. Tight control of target thickness helps narrow the final moisture distribution of your product.

Mills with poor thickness control experience excessive degrade and high sticker costs as well. The data in table 1.1 from a mill illustrates how bad things can be.

Table 1.1 ***A Mill with Poor Thickness Control***

Item	*# Samples*	*Thickness* *Avg.*	*Min.*	*Max.*	*Range*	*% Thin*
2 × 6 16′	48	1.834″	1.423″	2.030″	0.607″	31.3
2 × 6 18′	48	1.754″	1.385″	2.577″	1.192″	35.4
2 × 4 12′	51	1.745″	1.565″	1.914″	0.349″	25.5
2 × 8 12′	32	1.774″	1.575″	1.940″	0.365″	18.8
4 × 4 16′	20	3.841″	3.025″	3.976″	0.951″	15.0

This mill has serious drying problems (figure 1.2). Sticker losses are high, as stickers conform to the uneven surface across the width of stacked units. Thin boards side by side with thick boards dry without restraint and are prone to warp, contributing to degrade. Many do not dress at the planing mill. Thick pieces dry more slowly and force you to run kilns longer to assure that all the lumber is dry. They all increase machinery wear or require extra handling at the planing mill. Imagine the impact of that 2.577-inch thick board as it enters a planing machine at 1,000 fpm.

Thick lumber is a significant source of loss, yet it pales in comparison to thin lumber as a source of loss. Unless controlled, the mill in our example must increase target sizes to reduce the 15% to 35% thin production. This affects the very survival of any mill because it impacts on lumber recovery factors and grade yield. Although this is an extreme example, it illustrates the potential problems of variations in thickness. The problems do not go away with smaller variations; they simply reduce in magnitude. Variations

Figure 1.2 The extreme thickness variation at our example mill results in massive sticker breakage in addition to its obvious impact on drying and lumber quality.

Figure 1.3 Wide variation in board thickness allows thin boards adjacent to thick boards to move at will during drying. Stickers kink as they conform to the uneven surface and usually break the next time through the kiln.

in final moisture content continue. Machinery wear and additional handling costs are real. Stickers forced to conform to uneven lumber surfaces break the next time through the system (figure 1.3). Thin boards that dry without support are uneven in appearance and often lose grade or must be trimmed after dressing. Good thickness control is essential to quality drying.

Figure 1.4 Sawing taper into the lumber results in the sawtooth appearance of the tops of units waiting to be dried. In addition to the potential air loss from baffles that cannot close, the thicker ends will be slightly wetter than the opposite ends.

Sawing taper into lumber is a related thickness problem (figure 1.4). A board that is tapered is thicker on one end than on the other. Tapers ranging from 0.002 inch to 0.004 inch per foot are common. The thicker ends dry with slightly higher final moisture content. Thin ends are no problem as long as they dress at the planing mill, although they will be drier. Tapered lumber is easily spotted in lumber stacked to enter the kiln. One end of the stacked lumber pile is always higher than the other. From a distance, the tops of several units in a line have a sawtooth appearance. Stacking 50 layers, each with a taper of 0.004 inch per foot, gives an overall stacked unit height difference of 3.2 inches end to end in 16-foot lumber. Overhead baffles do not close the gaps on this lumber, resulting in some air loss over the units, which upsets air flows. The serious consequences occur when stick layers notice the lumber surface is uneven at the stacker. Rather than alert supervisors to correct the problem, they often double up stickers or turn them on edge at the low end. This gives them a level surface for laying stickers but does not give full support to the lumber. The lumber warps as it dries.

Occasional variations in thickness from snaking band saws result in the loss of individual boards (figure 1.5). These boards may simply be too thin and often cannot be trimmed. When too thick, losses extend to neighboring boards and to stickers. Keep in mind that each snaked board you see has a companion. Correct this problem by examining your saw filing and saw guide maintenance practices.

Figure 1.5 Variation in lumber thickness from snaking band saws.

Variations in Width

Variations in width within a single size (2 × 4s, 2 × 6s, etc.) have very little impact on drying. The real problem occurs when variations are so great or targets so wide that we must drop a board from the width of the unit.

The decision to drop a board is forced by stacker-width capacity. Full count layers that are too wide to fit between the stacker guides force the mill to drop one board from each unit. Rather than fix the problem by tightening target widths, mills often make adjustments farther along in the system. Tightening target sizes pays off in increased kiln capacity and often postpones the need to build more kilns. Table 1.2 illustrates the impact of lost boards on kiln capacity in 8-foot-wide units.

Table 1.2 *Comparison of Unit Widths*

8-foot-wide	*# Pieces per Layer*				
	One Board Dropped		*Full Width Capacity*		
Item	*# Pieces per Layer*	*Total Width*	*# Pieces per Layer*	*Total Width*	*Increased Capacity*
2 × 4	23	92″	24	96″	4.3%
2 × 6	15	90″	16	96″	6.7%
2 × 8	11	88″	12	96″	9.1%
2 × 10	9	90″	9	90″	0.0%
2 × 12	7	84″	8	96″	14.3%

Drying rates are affected by the total width of a stacked unit. When stacking to the total 8-foot width, all units but the 2 × 10s are a full 96 inches wide. Often mills notice that 2 × 10s mixed in with other items in a kiln are drier. This is due to the reduced wood volume across the unit and is an argument for drying 2 × 10s separately, although some mills have found they can mix 2 × 10s with 2 × 4s, since the 2 × 4s typically dry at a slightly faster rate than other items.

At mills that have lost one board in width, the variation is 8 inches among all items. This can be a significant source of moisture variation when these mills mix widths within a charge.

Variations in Length

Poor trimming affects the overall airflow distribution in a kiln by leaving openings through which air bypasses the lumber. Airflow naturally follows the path of least resistance. Poor trimming reduces the efficiency of the air delivery system and creates small imbalances in airflow uniformity (figure 1.6). Its impact on drying is often overlooked. The two trimming problems most often found are variations in length due to trim saw settings and mistrims.

From any trim saw, several combinations of saws produce any single length. When the maximum length produced by a mill is 16 feet, only one saw combination produces that length. However, 14-foot lumber can be produced by trimming either the near or far end. Trimming 4 feet from either end or 2 feet from both ends produces 12-foot lumber, and so on. If these saws are not set properly, the actual lengths of lumber produced from the various saw combinations are not equal. I have seen variations to 1 5/8 inches. These variations make it difficult to butt trucks tightly end to end within the kiln, particularly when combined with poor even ending. The resulting openings offer an avenue for air to short-circuit around the lumber.

The suggested saw settings in table 1.3 are measured from the even end stop to the far side of the saw from the stop on the trimmer. As shown in table 1.4, these settings give a maximum range in length for all combinations of .25 inch. This results in a neat, tight unit going into the kiln. The minimum over length values do not take into consideration that *these lengths will be shorter by one saw kerf.*

Mistrimming and missorting lumber are far more serious problems. The extra-long lengths stacked into units create large gaps between units. This is a sawmill problem and responsibility for its correction rests there. Repair problems when you find them (figure 1.7). Such problems include the failure of saws to drop, motor failure, loss of limit switches, and logic problems. Canadian-type trimmers must run at speeds that permit operators adequate

Figure 1.6 Poor end trimming results in units that will not butt together. The resulting air loss affects the drying efficiency of the entire kiln.

Table 1.3 ***Trimmer Saw Positions (in inches from even end line)***

0′ saw	*2′ saw*	*4′ saw*	*6′ saw*	*8′ saw*	*10′ saw*
0.75″	25.00″	49.25″	73.50″	98.00″	122.25″
12′ saw	*14′ saw*	*16′ saw*	*18′ saw*	*20′ saw*	
146.50″	170.75″	195.25″	219.50″	243.75″	

Table 1.4 ***Resulting Length Combination Comparison Data***

	Maximum Length	*Minimum Length*	*Length Range*	*Nominal Length*	*Minimum Overlength*
8′ combinations	97.25″	97.25″	0.00″	96.00″	1.25″
10′ combinations	121.75″	121.50″	0.25″	120.00″	1.50″
12′ combinations	146.00″	145.75″	0.25″	144.00″	1.75″
14′ combinations	170.25″	170.00″	0.25″	168.00″	2.00″
16′ combinations	194.50″	194.50″	0.00″	192.00″	2.50″
18′ combinations	218.75″	218.75″	0.00″	216.00″	2.75″
20′ combinations	243.00″	243.00″	0.00″	240.00″	3.00″

time to position boards before they enter the saws. It is inexcusable for problems to continue without making efforts to correct them. Ignoring them affects the entire drying process and endangers personnel who must load lumber into the kilns.

Remember: Airflow is upset by large gaps in the lumber. If mistrimming results in boards extending beyond the length of the units, then large gaps open between adjacent units on a truck. In severe cases, this limits the amount of lumber that will fit in the kiln, reducing kiln capacity. Once out of the sawmill, the problem can only be corrected by someone with a chain saw. When loading by forklift, individual units can be trimmed at ground level on the stacker roll case or before setting on the trucks. For transfer operations, there is no way personnel can trim this lumber and follow prescribed chain saw safety procedures.

If safety is not reason enough to trim it right in the first place, then consider money. The costs for manpower, fuel, and equipment are obvious. Are you absolutely certain that the end of the board that was trimmed is the end

Figure 1.7 Positioning the ragged ends of units next to the end baffles will help overcome the problems associated with the odd unit that is poorly trimmed.

that should have been trimmed? When trimming is this sloppy, the stacking crew pays less attention to even-ending lumber at the stacker. If not even-ended, the board may not have needed trimming in the first place. It will be trimmed again at the planing mill after adding on drying and handling costs. Did you needlessly lose 2 feet of board? Who knows? You can never be certain. When you consider how much one man can cost a mill with a simple chain saw, the decision to trim it right at the sawmill is easy (figure 1.8).

An unusual trimming problem occurred at a mill that dried studs from two company mills. Those that were shipped in were uniform in length and fully loaded the kilns. The local mill produced untrimmed studs cut from cores that were 8 inches longer. Since lumber from both mills could not be dried together, the local mill lost one kiln car from each track for a capacity reduction of 17%. A kiln charge of local lumber left a 6-foot gap on each track, which the end baffles could not close. This open avenue for air lowered overall air velocity and extended drying time. Moisture distributions were very wide. Determining when to pull the kilns was a guessing game you could not win. This was a design flaw at the plywood plant. There simply was no room for a trimmer.

Mismanufactured Lumber

Mismanufacture is any deviation from established target size. In practice, however, most people take this term to mean that the lumber has an unnatural defect manufactured into it. Snaking saws produce a type of mismanufacture we have already discussed. Skipped lumber resulting from mismatched saws in horizontal or vertical double arbor machine centers is another (figure 1.9). Rabbeted edges from Chip-n-Saws and related machinery are particularly noticeable (figure 1.10). These defects do not affect drying unless they are serious enough to alter the drying speed at the defect or damage lumber in their vicinity. The most serious considerations with these defects are those of appearance, grade, and volume yield.

The grading rules establish the amount of each defect permissible in each grade. Finish and shop grades are graded on appearance or suitability for remanufacturing. They do not permit these types of defects. Structural grades are judged by strength and often permit a significant amount.[4]

For example, a rabbeted edge is graded according to wane rules. On a #2 2×4 it may extend 1/2 inch deep and a little over 1 1/8 inches wide for the full length of the board. In 5% of pieces it may extend 1/2 inch deep and 1 3/4 inch wide or 1 inch deep and 7/8 inch wide for one-quarter the length. This material is difficult to sell in today's marketplace. You can never afford to overlook the importance of appearance and its impact on the customer.

Figure 1.8 Sawmill mistrims are dangerous to correct at the kilns as well as potentially expensive. When end trimming is this bad the stacker operator often fails to even-end the lumber. Does this board really need trimming, and if so, is he trimming the right end?

Figure 1.9 Mismatched saws on the double arbor gang usually result in objectionable skipped dressing down the board length.

It pays to take the position that you will not sell anything that you yourself would not buy. A high percentage of lumber goes into home projects built by homeowners shopping at the local builders' supply. Appearance is crucial. Think about it. When you shop at your building supply store for that weekend project, you are spending your money. That porch, deck, or shed is a reflection of your pride in ownership. If you are like me, you want the best you can afford. Appearance means a great deal in these decisions. The people who purchase your lumber feel the same way.

As a quality supervisor for the Southern Pine Inspection Bureau (SPIB), I inspected a carload of 2 × 4s delivered to a truss plant. This customer was proud of his trusses and objected to the skipped appearance of the lumber. Almost every board showed skipped dressing from mismatched saws. The lumber had been dried in a shavings-fired, direct-fired kiln, and the skips stood out under the yellow millbright coating. The load looked pretty rough, but the skips were not deep and were well within the rules. Five percent of a shipment may be below grade and, after inspecting a few bundles, I found the total degrade too close to recommend a complete claim inspection. The customer still did not want the lumber.

During a break I had the opportunity to watch the plant in action. Lumber from one supplier was breaking up as it passed through the truss saws. My suspicions aroused, I inspected a full bundle on the yard and found it

Figure 1.10 Rabbeted edge lumber. Is production so paramount that we will put our lack of pride on display for all to see?

13% below grade, mostly due to cross breaks. When I presented this information to the plant manager, he replied that this lumber was from his best supplier. It was always clean and good looking with little wane or warp. He paid a premium price and was glad to have it on his yard. He would not call a claim inspection. Appearance is important to your customers, too. Many will pay extra for it.

Mismanufacture that affects drying is rare and usually involves the surface quality of the lumber. When saws are not maintained or properly filed, a fuzzy surface results. This is a particular concern to mills processing steamed peeler cores from plywood plants. Dull sash gang saws often tear grain at board edges (figure 1.11). Both problems slow airflow through the stickers, extend drying time, and contribute to wide moisture distributions. I have seen situations where you could not see through sticker openings from one side to the other due to the fuzzy surface.

The quality of manufacture from any sawmill is a statement of the pride and responsibility of its employees. It cannot be shifted down the line to someone else. Quality of manufacture affects the total productivity for the mill. The log yard, sawmill, stacker, dry kilns, and planing mill all must work together to identify and solve problems. Everything you do—or don't do—has an impact farther along in the system. It is always easier to do a job right the first time than correct the problems of poor workmanship.

Figure 1.11 Fuzzy edges resulting from dull sash gang saws. This results in slow and uneven drying when it is severe.

CHAPTER GUIDELINES FOR REDUCING VARIABILITY IN DRYING

1. Rotate log inventory through the sawmill on a first-in/first-out basis to reduce predrying of logs on yard. Take control of your log inventory rather than letting it control you.
2. Dry lumber manufactured from water-stored logs separate from other production.
3. Implement a quality control (QC) program in the sawmill to reduce and maintain target sizes for thickness, width, and length.
4. Notify the people responsible when poor manufacture is observed. Repair the cause immediately.
5. Hold the sawmill personnel accountable for the quality of the lumber. This can be done by deducting plantwide waste from the sawmill cut.

REFERENCES

1. Denig, Joseph, Stephen J. Hanover, and C. Arthur Hart. *Kiln Drying Southern Yellow Pine.* The North Carolina Agricultural Extension Service. Publication No. AG-402, 1988.
2. Koch, Peter. *Utilization of the Southern Pines.* Agriculture Handbook No. 420, U.S. Department of Agriculture, Forest Service, 1972.
3. Brown, Terence D. *Quality Control in Lumber Manufacturing.* San Francisco, CA: Miller Freeman Publications, 1982.
4. Southern Pine Inspection Bureau. *Grading Rules—1977 Edition.* Pensacola, FL: Southern Pine Inspection Bureau, 1977.

Chapter Two

Sorting and Green Lumber Handling

A goal is a dream with a finish line.

Author unknown

Why Sort Your Lumber?

The primary purpose of sorting is to separate lumber items for selling or further processing. Sorting makes better use of drying capacity by allowing you to pack more lumber into your kilns. Sorting improves drying quality. It is rare to find a mill that does not sort directly out of the sawmill. For highest quality drying, each thickness must be sorted for length and width, then stacked and dried in charges of one thickness, if not width. This is the recommended practice at all sawmills.

Many years ago you could find mills that mixed all widths and lengths of a thickness into box piled packages for drying. The dried lumber was then sorted in a "rip mill" by width, length, and grade. Rip mills were equipped to upgrade lumber by ripping or trimming out growth, manufacturing, and drying defects. With cheap labor, this approach had distinct advantages, since it allowed mills to manufacture a wide range of specialty items in specific grades, including flooring, paneling, moulding, and partition. For example, such a mill might accept orders for 1 × 8 C&Btr paneling, pattern 139. With pregraded rough lumber, the finishing superintendent simply selects the required footage of 1 × 8 C&Btr rough stock and dresses it to pattern. Without this capability, a mill either runs a pattern from mill run and markets the downfall, or presorts before dressing. Due to the intense labor and equipment requirements, this practice is now uncommon in the South, although as mechanical sorting capacity becomes more affordable, mills are beginning to take another look at this option.

Figure 2.1 Drying lumber without the benefit of sorting. This mill mixed 4/4, 4/5, 6/4, and 7/4 lumber into random-length, box piled stacks for drying. The grade losses were beyond comprehension.

Years ago I visited a mill that had no sorting capability between the sawmill and kilns. They mixed all widths, lengths, and thicknesses for drying. Grade losses were beyond belief (figure 2.1).

How Should We Sort?

Here is the ideal way to sort: sort to solid, uniform unit size, separating items by length, width, and thickness. This results in maximum kiln capacity, highest grade recovery, and eases handling through the system.

Sorting for thickness is essential. Since drying time increases with thickness, this will always be an important component of quality drying. High-temperature drying of 4/4 takes anywhere from 8 to 16 hours. For 7/4 this range is 16 to 24 hours. Thinner lumber will always be overdry when two thicknesses are mixed.

Occasionally, I find a mill double-stacking 4/4 by placing two boards on top of each other within each layer. Complete units are then dried along with a 7/4 charge. On the surface this looks like an acceptable practice because it saves stickers and the lumber takes about the same amount of time to dry. Mills that are unable to cut sufficient 4/4 for a complete charge within a reasonable time may have no alternative. Keep in mind that, when dried, one side of each board is drier than the other and will warp (cup) after dressing, if it has not already done so. Some cupping may be acceptable in

Figure 2.2 Double-stacking 4/4 for drying. Although cupping may be manageable in 4-inch widths, it always causes problems in the wider widths.

common grades of 4-inch-wide stock. Six-inch and wider lumber cups severely before and after dressing when double-stacked for drying (figure 2.2). Reduce this cupping by giving rough dry stock sufficient time to equalize in the sheds before dressing, or equalize it within the kiln. However, neither of these options will remove cupping due to case-hardening; that requires conditioning in the kiln. Never dry lumber intended for finish or furniture grades in this way. Consider building a small kiln for these items if you want them dried properly, or you can run a double track kiln as a single track if you have the available kiln capacity,

At a few mills, reliable thickness sorting is a problem, or they may not saw enough 4/4 to justify separation. These mills often double-stack 4/4 within bundles of 7/4. Separation is done at the planing mill. Two 4/4 boards do not equal one 7/4 in thickness. In addition to cupping problems, expect excessive sticker loss and degraded 7/4 lumber adjacent to the double-stacked boards. If your mill cuts a full 2 inches (yes, a few still do), sticker losses and degrade are less concern, though warp remains. Always avoid mixing thicknesses in a kiln charge.

Separate widths for drying and run charges of one width if possible. An exception is in computerized multizone kilns using exiting air heat control. It is generally acceptable to mix widths in these kilns. However, keep in mind that computerized analysis of drying rates of different board widths indicates a 3% moisture difference when drying 2 × 4s and 2 × 10s together.[1]

Figure 2.3 Mixing widths in the same kiln charge. Endeavor to mix no more than two widths (4 inch and 6 inch, 6 inch and 8 inch, 8 inch and 10 inch, 10 inch and 12 inch) in a single charge, and place the wider widths on top.

We have already discussed how the differences between 2 × 4s and 2 × 10s are largely offset due to the differences in wood volume across the package width, so that it is often possible to dry these widths together. Do not mix 2 × 4s with any other item if you can help it. If you must mix, then only mix widths together that differ by no more than 2 inches (4 inch and 6 inch, 6 inch and 8 inch, 8 inch and 10 inch, 10 inch and 12 inch) (figure 2.3).

Mills that endeavor to avoid mixing widths into a kiln charge should be commended for their efforts. However, they should be aware that at certain times of the year this may be a hindrance. In the fall, when there are warm days and cool nights with low humidity, you may find that for items that take longer to accumulate (say, 2 × 10s and 2 × 12s) a significant amount of air drying will occur while you wait to get up a load. This increases the variability of the incoming moisture and widens moisture distributions. At these times, go ahead and mix widths to ensure that you have no more than three days between the oldest and youngest lumber going into the kiln. When the weather changes, go back to your regular practice.

If you set up kilns by forklift and dry several widths in the same charge, place the wider widths on top to take advantage of the increased airflow at that position.

Sort all lumber by length. Sorting each separate length is often a problem for a mill. Some mills were built with insufficient sort capability. More

Figure 2.4 Notice the distortion that occurs when too many lengths are stacked in a unit. Two lengths in the same unit is the maximum you should allow.

often, new product sorts were added to existing systems, increasing the need for sort bays. In this situation, a mill often sorts two or more lengths into a bin (figures 2.4 and 2.6). Never sort more than two lengths (8s and 10s, 10s and 12s, and so on) into one bin. If less than half the sorted lumber in a bin is the longer length, consider separating them or reassigning length combinations. Otherwise, expect high trim and sticker losses in the overhanging ends. When stacking this lumber always place the longest lengths on the outside edges of units. This helps maintain even airflow within the kiln and improves lumber support to reduce warp (figure 2.5).

Sorting to individual lengths increases average kiln capacity and is a modest means to improve kiln throughput. Add this to any justification to increase sort capacity. Lumber sorted to length is more uniform in final moisture content and will have a higher grade out. Units with mixed lengths have less lumber across the width on one end. This end dries faster, since the same amount of heat is spread across fewer boards. It overdries, causing warping, breaking of stickers, and degrade. Often the units have so few long lengths that end stickers are left out altogether (figure 2.7). These ends dry without restraint, moving at will. Trimming of split, warped, and shattered ends becomes accepted practice. When the monthly sawmill and planer production reports come across the manager's desk, does he ask, "Where did all the long lengths go?"

Figure 2.5 Mix no more than two lengths in any unit and always insert stickers. Stacking the longer lengths on the outside edge results in better support and causes less airflow disruption.

Figure 2.6 Mix no more than two lengths in any unit. This mill needs to re-evaluate its sorting strategy to eliminate huge holes like this. Sorting this poor impacts air flow patterns for the entire charge. All this for lumber ends that will likely be trimmed at the planer anyway.

Figure 2.7 If there are too few boards to insert stickers, unrestrained boards move at will as they overdry. This condition increases trim loss at the planing mill due to the overdry condition of the exposed board ends.

Make sure that enough boards are sorted into each sort bay to make complete units. In forklift operations, one bin load usually fills a single stacked unit. When there are too many boards, they are often so few in number that they are dropped out at the stacker. However, a slack stacker crew may simply create an extra partial layer, raising havoc when the unit is set up to load a kiln charge (figure 2.7). Unnoticed, top layers rake off as trucks are pushed in. If there are too few boards to fill the top layer, spread the boards out by spacing them evenly across the top. This distributes the weight of units placed on top, reducing compression damage and saving crossouts and stickers from breakage.

Things get a little trickier in transfer operations. Too many boards often amount to several layers. Dead pile them on crossouts and remove them with a forklift before loading the kiln (figure 2.8). Wide variations in stacked truck heights are not acceptable. An extra layer here, a missing layer or two there, and gaps develop under overhead baffles. These openings allow air to escape around the lumber, reducing overall air velocities within the kiln. Avoid more than two layer differences in height.

Never stack a unit unless you have sufficient lumber to complete the package. Starting a truck of 16-foot, only to finish it out with 14-foot, leaves a gaping hole for air loss. This has a negative impact on drying (figure 2.9).

Figure 2.8 Too much lumber in a stacked unit that has been dead piled on top for later forklift removal.

Figure 2.9 Loading your kilns with gaps results in reduced overall airflow through the lumber and represents a lost drying opportunity.

Sorting for Moisture Content

There is increasing interest and discussion of the merits of sorting for moisture content or wood density before drying. Just imagine how uniform final moisture distributions would be if we could narrow moisture distributions going in. We would shorten drying time, reduce degrade, and not worry about moisture kickbacks from the planing mill. On a practical basis this idea has not progressed too far in the South, although it is seen with increasing frequency at northern stud mills. While it is possible to make rough separations, this increases the need for sort capacity. Sorting two moistures doubles the required sort bays. Sorting three moistures triples them, and so on. The capital cost of sorting equipment and the extra handling it requires will limit its initial acceptance to mills with limited product lines, such as stud mills.

In northern climates and on certain species it often pays to sort out certain boards for special handling or drying. The first involves standing dead timber that is brought into the mill months, if not years, after it was killed. In many cases, the moisture content of this timber is already at or near acceptable levels, particularly when grade marking it S-Dry. To obtain a KD-19 stamp requires only a short trip through the kiln. If this timber is mixed with a normal green run and processed through a regular kiln charge, you will incur high degrade at the planing mill. It makes good sense to sort this material out of the normal production.

A second reason for sorting involves timber that is prone to water or wet pockets (balsam fir, eastern white pine, hemlock, etc.). Out west they call this timber sinker stock since it will sink in water if submerged. What makes this material so difficult to dry is the slowness with which the water in these pockets evaporates. Since the pockets contain solid water with no air, unlike the surrounding wood, the pockets dry from their outward surface inward, slowly diminishing in size. Imagine a puddle of water along the side of the road after a rainstorm. The puddle persists long after the pavement has dried. Fortunately, these wet pockets are readily visible on the board surface at the sawmill and can be pulled separately. It is common practice to set this lumber aside, stacked for air drying, for a month or two prior to entering the kiln to allow the pockets to dissipate.

Green Lumber Handling

There are two categories of green handling methods. One is found at mills that sort units for forklift handling. The other sorts for transfer cars.

Forklift unit size is limited by forklift weight capacities, and sorter bins are usually designed to hold one complete unit. The number of units required per kiln cart also depends on kiln door height and the maximum lumber length sawn. A mill might stack two units for 12-foot and under with three for everything else. Five units may be required for tall doors and 20-foot and longer lumber. I recommend keeping the number of units per cart uniform to reduce variations in airflow due to varying numbers of crossouts (see table 4.2). Use the fewest units possible for your operation. As the table illustrates, a lot of air escapes through the crossouts. Many mills are built with transfer cars because they can stack an entire kiln cart without crossouts. Others look for ways to reduce or eliminate them. Budgeting for forklift replacement this year? Consider higher capacity forklifts to reduce crossouts, keeping in mind sorter bin capacity.

A great deal of green end handling equipment is built too light for green end use. If the equipment is a chronic maintenance problem, investigate replacing it. As you compare the equipment from several suppliers, compare the total weight of steel in each. Look at the horsepower. Low initial purchase prices are soon offset by downtime and repair costs. Go for rugged, overdesigned equipment when considering sorting and green handling equipment. You will not regret the investment.

Storing Green Lumber in Windrows

Lumber is often simply scooped up loose at the sorter by forklift and transported to long windrows for temporary storage prior to stacking (figure 2.10). These windrows are organized by laying down wooden or concrete

Figure 2.10 Green lumber held in windrow bundles prior to stacking. Never pile bundles on top of each other due to the breakage you will incur.

Figure 2.11 Green lumber held in bundles prior to stacking. Keep bundled lumber off the ground and separated so it can be retrieved with a minimum of damage. Two binders would help keep this lumber from crossing up.

Figure 2.12 Green lumber bundles stored with one end resting on the ground. The ends pick up rocks and grit, which hastens knife wear at the planer.

sills designed to keep lumber off the ground (figure 2.11). Often it does not. Individual sort bundles may be banded or chained into a loose unit. This increases handling costs but reduces downfall between the sawmill and kilns. Even here, loose and broken boards pile up and are lost from the system. Ends lie on pavement or in mud, picking up rocks and grit that reduce planer knife life (figure 2.12). These operations often have an unkempt appearance and high downfall factors. They also have more planer downtime for knife changes.

Figure 2.13 A well thought out windrow storage system. Lumber is well supported on clean, solid supports and is accessible from either side to keep it rotated on a first-in/first-out basis.

To reduce losses from running a windrow storage system, pave the storage area with concrete or asphalt. Simple graded and graveled yards are not acceptable. These yards rut and are a ready source of rocks and grit for your planer. Provide adequate drainage. Keep the area clean, sweeping occasionally and removing trash and broken boards. Align forklift alleys north/south to keep them dry in winter. Position two sills for each row. One sill is usually not adequate unless it is very broad. One end of the pile always seems to rest on the ground. Space sills one-half the length of the lumber planned for that row. For example, rest 16-foot lumber on sills 8 feet apart or wide enough for the lift to pass. Give yourself plenty of walking room (3 to 4 feet) around the lumber ends for inventory and inspection. You will appreciate it later. Anchor sills by pinning them to the ground with rebar rods. The sills may be wood, concrete, or steel, 6 inches thick and at least 6 inches wide—the wider the better. Do not use poles. At the sorter place two bands or chains around each sorted bundle. Never move lumber loose or with one band. This prevents lumber within the bundles from crossing up and breaking. Store bundles one deep to prevent breakage. Make each row accessible from both ends to allow the lift driver to place fresh green lumber on one end of the row and stack older lumber off the other side (figure 2.13). Otherwise, the last bundle in a row always seems to hang around too long. Stacking and placing lumber into the kiln as soon as possible after manufacture gives us highest lumber quality.

Transfer operations stack entire kiln carts at the stacker. Lumber sorted for these operations is held on a green yard until there is sufficient lumber to stack a cart. If lumber is held in windrows, setting up the green yard follows the same rules as a forklift yard,

Storing Green Lumber in Dead Piled Units

Some mills run two stackers, with the second one at the sorter. The contents of a sort bin are stacked into units without stickers (dead piled) and banded before transfer to the green yard. This method saves yard space and

Figure 2.14 Green lumber held in dead piled units. Note how every other layer is offset on one end to knit the package together. This avoids the worry of dealing with strapping systems.

Figure 2.15 Green lumber held on stickers. Note the lumber distortion that results from insufficient ground support and misplaced crossouts.

reduces degrade when compared to windrow storage, but requires more labor. The green storage yard is set up the same as for stickered lumber in a forklift operation. Want to save the labor and expense of banding? Try this method in use at several mills: As successive layers are deposited by the stacker, a small air cylinder automatically pushes every other layer over an extra few inches on the even end. Since boards do not line up one above the other in the package, the unit "knits" together and can be moved about without bands (figure 2.14).

The most efficient forklift operations operate with the lumber stacker direct coupled to the sorter. This forces a higher level of maintenance on the sorter and stacker, since unscheduled downtime eventually shuts down the sawmill. When people are expected to operate at a higher level, they develop the attitudes necessary to get the job done. These attitudes carry over throughout the mill.

Direct coupling is the least expensive alternative. By forcing the mill to maintain its equipment at a higher level, the entire sawmill runs better. Design in additional surge capacity or ways to off-load and introduce lumber prior to the stacker if you feel you need it. The extra handling costs required to move green lumber in and out of storage prior to stacking are eliminated. Grade and volume loss is held to a minimum. All lumber is stacked and held on stickers to get a leg up on drying.

Storing Green Lumber on Stickers

Stacked lumber must be held in storage before placing it into the kiln (figure 2.15). Grade and pave this storage area to give the forklift a level working surface. Rough, rutted roads cause boards to fall out and stickers to shift, degrading lumber. Provide good drainage. Align alleys north/south to keep them dry. Stickered lumber is flexible and requires good support. Oak or treated timber sills, 6 × 8 inches and larger, make good foundations when arranged to line up with stickers. Place two sills under lengths to 10 feet, three under lengths to 16 feet, and four under everything else. Add additional sills if you see bottom layers sagging. To reduce crossout handling, use crossouts from the kiln. I recommend 4 × 4s located at each sticker location. Some mills use 2 × 4s on edge for crossouts, but I find they shift easily and fall over. They do not last as long, either. A few mills have found that they can use 2 × 4s laid flat in conjunction with thin forks on the forklift. This is a good idea for reducing air volume loss through the crossouts and perhaps gaining a little kiln capacity. The dried 2 × 4 crossouts are then simply fed through the planer with regular 2 × 4 production.

Always keep your lumber as even-aged as possible by stacking the oldest stock first. This avoids an important cause of spotty wet lumber (figure 2.16). One week is the limit for holding lumber before stacking. Fresh lumber mixed with lumber that is over a week old is difficult to dry to uniform moisture content (figure 2.17). At certain times of the year five to seven days may be too long.

Figure 2.16 Mixing green lumber with lumber that is seriously blue stained invites moisture complaints from the planing mill.

Figure 2.17 Something is wrong with the green inventory system to allow this type of age or condition mixing. This blue stained unit will be harder to dry than the fresh lumber around it and will likely result in wet lumber at the planing mill.

Figure 2.18 Good sorting and handling improve your chances of avoiding drying problems.

Most transfer operations move entire loaded kiln carts from the stacker through the kilns to the planer breakdown. Some insert crossouts for fork-lift handling behind the kiln.

Compared to a well-run forklift operation, transfer operations are labor and equipment intensive. Nevertheless, they seem to be favored by the

Figure 2.19 Even subtle differences in lumber dryness as it enters the kiln show up after drying.

larger mills. Transfer cars endure much abuse, and breakdowns are frequent. Cables used to move cars and pull kilns often break or slip off, causing injury. I once watched amazed as a cable suddenly broke and caught a man's ankle, turning him head over heels. He was fortunate—only the cable was broken. Cart wrecks, in and out of kilns, are more frequent. Kiln trucks also sustain more damage and incur higher repair costs.

Whichever system you use, place all lumber into the kiln as soon as possible to reduce surface drying and fungal degrade (figure 2.18). Fresh green lumber is always easier to dry. Lumber held on the yard long enough to blue stain takes a little longer to dry. However, lumber that is blued and a month or more older takes less time. What is important is that each kiln charge be composed of lumber of a similar age. This narrows final moisture distributions and improves grade (figure 2.19).

Setting Up an Air Drying Yard

From time to time and for a variety of reasons it often becomes necessary for mills to set up a temporary air drying yard. The rules for temporary yards are similar to for permanent yards.

When stacking lumber for air drying it is common practice to space the boards out on each layer to allow air movement down through the packages. In 6-foot-wide packages and wider it is a good idea to build a central chimney into the package so that cool moist air can flow down through the center of the unit, drawing warm, dry air into the sides. This will speed drying.

The storage yard is laid out so that the prevailing wind blows parallel to the length of the units, not parallel to the stickers, as is popularly believed. The mechanism for air movement on an air drying yard is not analogous to a kiln where we try to force air directly parallel to the stickers. If we set up our yards in this manner, all of the energy of the prevailing wind is dissipated in the first stack of lumber it meets. The first stack will dry great. However, as we travel deeper into the rows of lumber the air becomes stagnant, and drying slows. The purpose of the wind in an air drying yard is to sweep away the cool, moist air which falls to the ground around the stacks so that warm, dry air will be drawn into the top of the packages to replace that which filters down. This results in the fastest, most uniform air drying.

Build solid, level supports in an area with good drainage that takes advantage of the prevailing winds coming off sun-soaked parking lots, highways, or other pavement. Place a support under each sticker location to avoid warping. Space stacks out at least 3 feet in every direction to maximize air movement, and keep the lines of stacks in a neat row. Keep the weeds down by mowing or spraying, and do not forget to mark each package with the date it was placed on the yard. If occasional air drying looks like it will become a continuing practice, you might consider building some simple roofs using low-grade lumber that can be placed on stack tops by forklift to protect the top layers from the sun. Contrary to popular belief, air dried lumber is not superior to kiln dried, since it lacks a critical element for good drying—control of the drying environment. Over the years I have helped many mills make the transition from air drying to kiln drying, and they are always amazed at how much their grade improves once they complete the process.

CHAPTER GUIDELINES FOR REDUCING VARIABILITY IN DRYING

Sorting

1. Sort and dry lumber by thickness. Never mix thicknesses within the same kiln charge.
2. Sort and dry lumber by width. If you must mix widths within a kiln, they should vary by no more than 2 inches.
3. Sort lumber by length. If you must mix lengths in a unit, mix no more than two lengths, and place longer lengths on the outside.
4. Sort enough boards into a sort bin to complete units.
5. Never mix package lengths on a kiln cart. This creates large gaps for air loss.
6. Build bundles that are uniform in height and width.

Green Lumber Handling

Forklift Operations

1. Use the fewest number of units possible per kiln cart to reduce crossout openings. Keep the number of crossouts constant between carts.
2. If possible, run the stacker direct coupled to the sorter. Avoid handling unstacked lumber.
3. If you must yard green lumber before stacking, dead pile into units or band into bundles. Use two bands or chains per unit.
4. Remove excess boards from the system; do not add extra layers. Space boards in unfilled layers across the package width.
5. When storing stacked green lumber prior to placing in kiln, place a crossout at each sticker location. Place lumber on sufficiently level and aligned sills to prevent the bottom layers from sagging.

Transfer Operations

1. When yarding green lumber for stacking, dead pile into units or band into bundles. Use two bands or chains per unit.
2. Completed stacks should be the same height. Avoid variations in stack height that exceed two layers.

For Both Systems

1. Stack and place lumber into the kilns as soon as possible after manufacture. Lumber in any kiln charge must be as even-aged as possible to reduce drying problems.
2. Plan the loading of each kiln charge rather than letting it happen!

Guidelines for Establishing a Green Handling Yard

1. Grade and pave area for a level and dry surface. Provide good drainage. Keep area clean and free of debris. Align alleys north/south so the sun can dry the pavement and melt snow.
2. Position two support sills for each row, spacing them one-half the length of the lumber they will support. Anchor them.
3. Sills may be rot-resistant wood, concrete, or steel, 6 × 8 inches or larger.
4. Space adjacent rows of lumber with 3 to 4 feet between for inventory access.
5. Make each row accessible from each end to keep stock rotated on a first-in/first-out basis.
6. Band each bundle with two bands, one on each end about 2 feet back from the end. Do not pile bundles on top of each other.

Guidelines for Establishing a Stacked Green Lumber Yard

1. Grade and pave the yard, and provide good drainage. Align alleys north/south. Keep the yard clean.
2. Use sills made from rot-resistant wood, concrete, or steel, spacing them to line up with stickers and eliminate sagging.
3. Place crossouts at each sticker location or at least at the sticker where the sill is positioned.

Setting Up an Air Drying Yard

1. Align stacks with the prevailing wind parallel to the length of the lumber. If this is not possible due to the terrain, then align the main forklift alleys north/south to keep them free of snow and ice in winter.
2. Choose an area with good wind circulation that takes advantage of any heated air coming off adjacent pavement.
3. Install solid supports that are level in all directions under each sticker. The higher the stacks are off the ground, the better.
4. Space adjacent stacks no closer than 3 feet in any direction.
5. Use portable roofs on each stack to protect the top layers from the sun.
6. Kill the weeds or keep them trimmed back.
7. Label each unit as it is placed on the yard so you know how long it has been there.

REFERENCES

1. Denig, Joseph, Stephen J. Hanover, and C. Arthur Hart. *Kiln Drying Southern Yellow Pine Lumber.* The North Carolina Agricultural Extension Service. Publication No. AG-402, 1988.
2. *Air Drying of Lumber—A Guide to Industry Practices.* Agricultural Handbook No. 402, U.S. Department of Agriculture, Forest Service, 1971.

Chapter Three

Lumber Stacking and Stick Laying

You may be disappointed if you fail,
but you are doomed if you don't try.
Beverly Sills

Improper stacking and stick laying is often the greatest source of degrade in a drying operation (figure 3.1). To understand why, we must visualize what happens to lumber during the drying process.

In your dry kiln, lumber goes through a "plastic" stage during the high heat and humidity of warm-up. The conditions are similar to a method cabinetmakers use to bend special wooden shapes for chairs and furniture parts. To form the arm of a bentwood rocker, the craftsman selects dry, straight-grained lumber and machines it to the desired cross-sectional shape. He then places it into a steaming chamber until the wood becomes pliable. After it is removed from the hot, moist environment of the chamber he bends and clamps the wood into a form, which determines its shape. Once the arm has cooled, he releases it from the form and works it into the chair. The wood will keep its shape.

The same process occurs in the dry kiln on a grand scale, particularly in the higher temperature drying schedules. Misplaced or missing stickers and those that vary in size do not hold lumber flat, and wherever there are variations in support the lumber will bend to conform to the forces placed on it. Once dried, the distortions are permanent. This shows up in dressed lumber as short skipped areas on the wide face or localized kinks along the length of the board (figure 3.2).

Correction of poor stacking methods is easy, but it is a continuing effort. Don't let up. The rewards are there. Good stacking becomes natural and

Figure 3.1 Poor sticker quality and placement degrades lumber. Note the localized distortions from thin or missing stickers and those laid on edge.

automatic as personnel get into the habit of doing it right. When training new people, explain *why* they are stacking lumber the right way as well as demonstrating *how.* Take them to the rough dry sheds or to the planing mill to show them the impact of poor stick laying. Most people want to do a good job, and will—if they understand that what they are doing is important to the survival of the mill.

Figure 3.2 Poor stick laying results in an uneven surface appearance, which is readily apparent at the planing mill. Short skipped areas in the vicinity of stickers are common. You can see all of your stacking problems at the planing mill. Spend time there every day.

Sticker Material

An amazing variety of sticker materials are available. What is most economical for your operation depends not only on initial purchase price but also on the factors that affect sticker life. Do wide variations in lumber thickness distort and break stickers? Do you handle stacked lumber by tram or forklift? Does sticker handling and storage equipment cause breakage? All these factors and more affect the actual cost of stickers. Determine which sticker is best for you by testing different ones (figure 3.3).

Estimate actual sticker costs by dividing the cost of a sticker by the average number of times it passes through the kilns. This is best determined by color coding the ends of a new unit of stickers and tracking them through the system. After each pass through the kiln, count them. When half remain, you know your average sticker life in charges per sticker. Looked at in this manner, an inexpensive sticker with a short life costs more than you might think. While conducting this test, try to determine why your stickers break. Is it their quality or are you doing something that shortens their life? Mismanufactured lumber, mixed lengths within stacked units, poor sticker handling, and other factors all affect sticker life. Each can be corrected.

The following stickers are commonly available:

Self-Produced or Purchased Domestic Hardwood

Hardwood stickers are highly variable in quality unless kiln dried to 10% moisture content before sizing. All that I have seen were green when

Figure 3.3 This mill has a variety of stickers in the system—LVL, apatong, perhaps a few softwood. As long as they are similar in thickness they cause no problem.

first used to stack lumber. Wood shrinks when dried, and shrinks at different rates depending on such factors as species, green moisture content, and the orientation of growth rings. Add variation in sawn green dimension and the final thickness of dry stickers in service is remarkable. This variation distorts the lumber it supports. The actual service life of hardwood stickers depends on the grade of wood they are made from. Grain variation around knots shortens their life. Tapered wane causes warp. Clear, straight-grained, dry hardwood stickers are prohibitive in price due to their value as clear lumber. Still, many mills continue to use green hardwood stickers, perhaps lulled by a low initial purchase price. The actual cost of these stickers is very high due to poor lumber support and short sticker life.

If you insist on using green hardwood stickers, be very careful. Inspect each shipment for uniformity and turn back loads with poor thickness control and excessive wane. Place them immediately into the system to get their moisture content down. Remember, they are green and subject to mold and decay. I have seen mills hold these stickers for months before placing them into the system. By then, decay had shortened their life to one charge. I never recommend these stickers (figure 3.4).

Self-Produced or Purchased Softwood

Softwood stickers are often produced at the mill from dry stock specially graded out at the planer. Graders select clear 2 × 8s and wider that have economy or saddle wane. Stickers are cut using cutoff and gang rip saws.

Figure 3.4 Hardwood stickers held in the original units from the supplier until they are absolutely worthless. If you insist on using green hardwood stickers, place them into the kiln immediately to prevent deterioration.

High-quality #1 dimension lumber is good if the need arises and the market is slow. All of these stickers are straight grained and have few knots to limit their life expectancy. They make good stickers, although you may notice a tendency for them to become thinner with use. Never make stickers from #2 common and lower grades—the knots and cross grain produce too much breakage. When you consider the initial material and manufacturing costs, these stickers can also be pretty expensive.

Phenolic Resin or Plastic Stickers

Occasionally you will find phenolic resin or plastic stickers, although their use is rare. I know of no one who uses them exclusively. Some concerns include the use of petroleum-based products for their manufacture; the disposal or recycling of broken stickers; and the tendency for plastics to become brittle under continuous high temperatures. One sticker company reportedly buys back broken stickers at about 25% of the purchase price and recycles them. I see these stickers predominantly at treating plants, which use them because they do not absorb chemical from their KDAT (Kiln Dried After Treatment) lumber.

Aluminum Stickers

At least one mill has made trial runs using extruded aluminum stickers. These stickers represent a significant investment that would be economical were it not for their tendency to bend rather than flex under load. Once bent, their life expectancy is very short. The primary concern governing their use is the recycling of bent and broken stickers. They are also hot to handle if not allowed to cool between uses, and over time they are reported to develop sharp burrs, which cut stick layers' hands.

Tubular Steel Stickers

This interesting variation on the aluminum sticker seems to have caught on at a few mills. These stickers also bend under load, but seem to be a little easier to use. Broken stickers are welded together into longer lengths and cut to normal sticker length. The mills that use them, like them.

Laminated Veneer Lumber (LVL) Stickers

LVL stickers are becoming more common as plywood plants seek to diversify their product mix. In general, these stickers are very good. There are two main types. The first has its layers laid flat across the width of the sticker. They give good service. A better LVL sticker has its layers laid flat on the narrow edge, showing glue lines on the wide face. This sticker is stronger and gives longer service life. All are uniform in size and dry when purchased. They also tend to keep their dimensions better and warp less than solid wood stickers. Most are made of softwood veneer, although occasionally they are made using hardwood. I have heard complaints of delamination or separation of layers in these stickers. The manufacturer should replace them when you notify them.

Apitong Stickers

These imported hardwood stickers are good for mills with high sticker losses resulting from poor lumber size control. They are highly durable and last a long time under severe service conditions. They are uniform in size and keep their shape well. I generally equate them with LVL stickers, although at many mills Apitong is considered superior. Western mills have used them for many years with good results. Their main drawback is the length of time it takes to get them once you place an order, although they have improved in availability over the last decade. Apitong is a high-density hardwood, which makes these stickers a bit heavier than most. You may hear some grumbling among stick layers who are used to handling lighter stickers.

OSB Stickers (Oriented Strand Board Stickers)

Do not go there.

Sticker Size

Thickness

Thickness is the most important dimension in a sticker, since it determines the spacing for each layer. Thickness must be uniform to provide even support to lumber without degrade. There must be free flow of air, since that is the reason for stickers in the first place. Mills that produce their own stickers must monitor their manufacture to assure quality. Incoming purchased stickers must be spot-checked and rejected if poorly made or the wrong size. Quality is often variable.

Use the thinnest sticker that gives good service life and the best drying quality. For every 0.125 inch reduction in sticker thickness you will gain an average of 5% in kiln capacity. Be careful here. Increases in kiln capacity are easily offset by high sticker losses, increased lumber degrade, extended drying time, and wider moisture distributions.

Most mills use $\frac{3}{4}$-inch or $\frac{7}{8}$-inch stickers and enjoy long sticker life—provided they have good lumber size control. Studies show a negligible impact on drying time between these thicknesses, although I personally prefer $\frac{7}{8}$-inch stickers.[1] It is interesting to note that as you reduce sticker thickness within many kilns, air velocity increases. However, since thinner stickers increase total lumber volume, any benefit to drying time is often offset by the increased heat demand. If your fan system is already performing near its maximum static air pressure capacity, air velocity may not increase.

A few mills use $1\frac{1}{8}$-inch- to $1\frac{1}{4}$-inch-thick stickers. They pay a steep price in lost kiln capacity. Many of these mills do this to extend sticker life and improve airflow due to poor lumber size control. Others have kilns with marginal heat capacity and seek to improve drying speed and lumber quality. Mills that dry a large proportion of timbers benefit from the extended life of thicker stickers for these items. If your fan system is undersized, thicker stickers tend to give more uniform drying across the load.

Studies done at Oregon State indicate that a sticker thickness as small as $\frac{7}{16}$ inch is practical for conventional kilns.[2] I question whether this research will transfer to the real world. At least one mill study has been done comparing $\frac{5}{8}$-inch stickers to $\frac{3}{4}$-inch stickers in a high-temperature kiln. The $\frac{5}{8}$-inch stickers do not perform as well. My primary concern about the thinner stickers is their service life, since they are intolerant of variations in lumber thickness.

Width

Stickers must be wide enough to assure that stick layers recognize the difference between width and thickness at normal stacking rates. One-and-a-half inches is a common width. Square stickers are acceptable in $1\frac{1}{8}$-inch to $1\frac{1}{4}$-inch sizes, but avoid smaller ones due to reduced sticker life. Tubular

Figure 3.5 Using short stickers that allow edge boards to drop down and close openings. Finding random whole layers of wet lumber at the planing mill? This could be the cause.

steel stickers are commonly 1 inch by 1 inch. Small Apitong squares reportedly work well also. Small square stickers do have a tendency to indent the surface of the wood if they are only slightly out of line, so be careful. Some stick laying machines require stickers 2½ inches to 3½ inches wide to work well and are usually manufactured under a Canadian influence. These stickers may be good for the machine, but they are bad for drying. Expect wet spots under these stickers even if the stickers start out dry. Many mills use green 1×3 and 1×4 material as stickers in these machines (figure 3.12). Talk about a wet spot. Do not purchase a stick laying machine that requires wide stickers to function reliably.

Figure 3.6 Short-term fix for the problem in figure 3.5. Alternate pushing and pulling stickers to the side to support lumber along its length.

Length

Stickers must be long enough to extend completely across the stacked unit plus a little extra. Never accept stickers that are much shorter than the unit width. Unsupported boards on the edges will drop down, covering the airspace, or simply warp in place (figure 3.5). If you get short stickers by mistake and must live with them, push the first sticker to the far side, pull the second one to the near, push the third to the far, and so on. Although the boards are not fully supported, at least they will not fall and block air circulation (figure 3.6).

Figure 3.7 Typical problems encountered with stickers as they age through the system. From left to right, crooked sticker, broken sticker, tapered sticker, too thick sticker, good sticker, thin sticker, thick sticker. All but the good stickers must be removed from the system.

Figure 3.8 A mill with no organized plan for sticker storage. Stickers represent a significant investment for any mill. Protect it.

Uniformity

Manufacture all stickers to uniform dimensions for even support but remember that the most critical dimension is thickness. This is often a problem with purchased hardwood or softwood stickers. Be cautious and inspect each truckload. Reject sticker bundles that vary in thickness, or are waney or knotty. Stickers are expensive enough without the added cost of breakage before use, or with variations that degrade your lumber. Avoid mixing different types of stickers in the system for the same reason (figures 3.7 and 3.8).

Stack Width

In terms of drying speed and narrowed moisture distributions across the width of a unit, the maximum stack width is around 8 feet. Stack widths of 6 and 7 feet are encountered, which helps narrow moisture distributions still further. However, this reduces the amount of lumber you can place in the kiln. If you are stacking narrow widths and need additional kiln capacity, consider going to 8-foot units. Keep in mind that you may have to beef up the heat and/or airflow systems to keep from extending drying time. Still, this will probably be much less expensive than building a new kiln. Many West Coast kilns and a few southern kilns built under West Coast influence are often designed to handle widths from 9½ to 10 feet. The wider units are only acceptable for slow-dried species, and I do not recommend them there. Unless these kilns have sufficient air velocity to overcome the large temperature drops as air passes through the lumber, moisture variation across the unit is large, and drying times are extended. This tendency can be overcome to some degree by using thicker stickers, but this reduces kiln capacity—the purpose of the wider width packages in the first place.

Mills making the transition from air drying to kiln drying often keep their narrow stack widths of 4 to 5 feet and set two bundles side by side on a truck. This is acceptable as a transition, but begin making plans to convert to the wider single unit. When two bundles are placed side by side on a single truck, sticker courses do not always line up. If the stacks are too close, wet spots appear in your lumber. Space units 2 to 3 inches apart on the carts to allow airflow to reorient from one stack to the next. If you are committed to using the narrower packages, use 2 × 4s stood up in pockets on the center line of the carts to maintain a space. Keep in mind that maintaining this stacking method generally results in twice as many units, which increases handling costs. The additional handling, in turn, introduces more opportunities for lumber damage and degrade.

Stick Laying

One cannot overemphasize the importance of proper stick laying. The full cost of untrained or inattentive stick layers is far more than wages, since it extends to the reduced quality of your lumber. Take the time to show new stick layers why their job is important; it is not enough to simply tell them. Take them to the dry sheds behind the kilns and point out the distortions caused by poor stick laying. Most people take a different outlook toward their efforts when they realize how poor workmanship on their part affects the mill.

Sticker Quality

Maintaining sticker quality is a continuing process. Repeated cycles through the kiln and handling equipment cause stickers to warp and break, eventually damaging the lumber they support. Remove any sticker with the potential to cause degrade from the system.

Most mills remove stickers at the stacker, relying on the stick layers to remove bad stickers. High stacking rates make this very difficult, since there is little time between cycles. If you depend on your stick layers to cull stickers, provide a method for sticker disposal. Many mills simply position dumpsters near the stacker. Others convey stickers to a nearby hog. The worst practice is throwing them out on the ground for later disposal. This invites fire and creates hazardous working conditions. The best system I have seen removes stickers at the planing mill (figure 3.9). As stickers pass by from the

Figure 3.9 Sorting stickers at the planing mill. Bad stickers are thrown onto the belt on the right and conveyed to the trimmer hog.

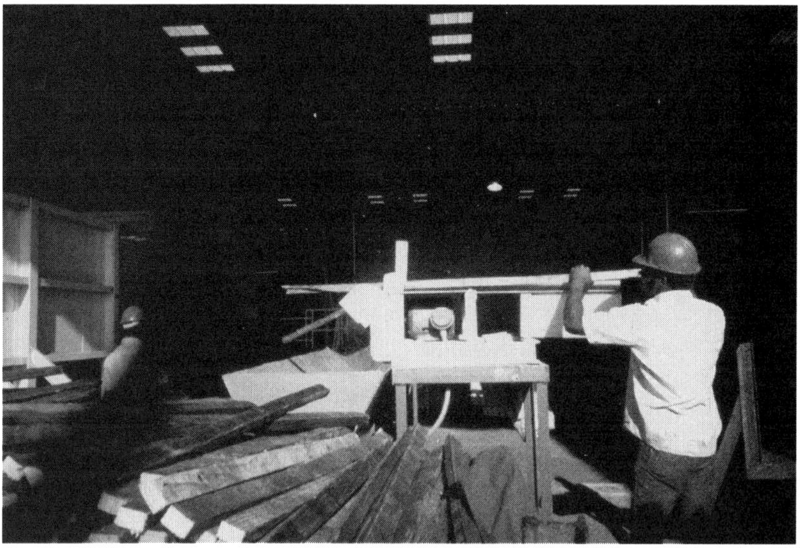

Figure 3.10 Culling and trimming bad stickers at the unstacker to reuse them as spacers in dressed lumber units.

unstacker, a worker inspects and discards bad stickers into a bin. He may drop them onto a conveyor to the trim saw hog. Some mills cut them into shorter lengths to use as spacers in finished lumber units (figure 3.10). This is a very practical option for operations that bundle stickers going back to the stick layer.

Stickers that are uniform in thickness do not cause degrade unless misplaced. Try to keep sticker thickness within a range of 1/8 inch. More variation than this results in small distortions that ruin the flat appearance of your lumber. This is important to your customer. Extreme variations degrade lumber and cost you yield. Remove all thick and thin stickers from the system.

Discard broken stickers that do not reach across the full width of your units. *These stickers cost more in one trip through the kiln than their replacement cost.* Unsupported edge boards warp at will. Boards drop down and cover the sticker space. The reduced air flows widen moisture distributions. If you are in a bind for stickers, lay two stickers side by side to extend across the width. Consider this a temporary measure only, as improper placement impedes airflow and slows down the stick layer (figure 3.11).

Crooked stickers upset air flows and produce wet and overdry spots in your lumber. On the outside edge of a curved sticker, air speeds up due to the restriction, and drying is faster. The inside of the curve creates a dead air space, which slows drying. The resulting moisture variations down the

Figure 3.11 Piecing short stickers together to extend across the package width. Consider this a temporary measure only, since alignment is critical and piecing slows down the stacker.

length of a board cause quality problems for customers using your lumber for furniture or laminated beams. Have you ever noticed how moisture contents can vary down the length of a board? Here is the prime reason for its occurrence.

Figure 3.12 Green 2 x 4s used as stickers at a stud mill. This is a poor practice that results in wet spots at the point of contact. Even dry 2 x 4s result in wet spots. As you improve your drying process and learn to dry faster, these wet spots will become worse.

Sticker Placement

Lay straight, uniform stickers perpendicular to the length of the unit, spaced on 2-foot centers (figure 3.13). Some mills stack on 4-foot centers or stack only the bottom 10 or so courses on 2-foot centers, with the balance on 4-foot centers. They maintain that their studies indicate no additional

Figure 3.13 Good sticker orientation and alignment. Unless straight stickers are always laid perpendicular to the package length, expect to find random wet spots down the lumber length.

degrade from this practice. One thing is certain: Sticker alignment becomes more critical with fewer stickers. I tend toward the conservative in this area and recommend 2-foot centers for all stickers and truck supports. Always use sticker guides that align with your kiln truck supports. Make a scale drawing to determine accurate sticker positions. This also allows you to determine spacing for other equipment that is affected by sticker spacing such as forklift forks, cart supports, and planer breakdowns.

Good sticker guides are sometimes hard to keep in service. The best guides form pockets on both sides of the load into which the stick layer drops his stickers. For many stackers this is mechanically impossible on the far side. Alternatives include hanging short, heavy chains from a rugged support to mark sticker locations. Lasers and shadow lines also work well, but be careful with shadow lines. The high temperatures that older ones produce may cause a fire if the area is not kept clean.

Mechanical stick layers are becoming common, but be very careful when considering their purchase. Look for rugged but simple equipment. Equipment with a lot of small working parts to maintain is difficult to keep in peak performance operation. The best ones I have seen place each sticker rather than drop them from a hopper. All require high sticker quality, but some have problems laying stickers consistently right. The most common problems I see are missing stickers and stickers laid on edge. Both cause degrade. Evaluate the available equipment very carefully to make certain the unit you buy does not cost you down the road. Visit mills running the equipment and watch it work. Walk through the stacked green lumber yard

Figure 3.14 Painting the ends of stickers is a good method for evaluating stick layer performance. The stick layer on the left either needs additional coaching or another job.

and look for missing stickers and those turned on edge or misplaced. Finally, ask the manager if he noticed any grade improvement after installing the mechanical stick layer. A good one actually improves grade recovery. After years of hesitancy on my part in accepting these machines, I think several of the manufacturers have finally got it right.

Lay stickers one above the other for continuous support from top to bottom in the unit. Unaligned stickers kink the wood, which show up as small localized skips at the planing mill. Check sticker alignment by spray painting a line connecting each sticker top to bottom in a unit (figure 3.14). Properly stacked units have straight lines. This method helps identify stick layers who need improvement. Paint the sides of a stacked unit and place it in clear view of the stick layers. You need not say anything to them. Sooner or later one of them will come to you wondering what you are doing. Reply that you are checking sticker alignment and demonstrate the impact of poor stacking practices. The improvement may surprise you.

This is an effective way to deal with stick layers who are slow to learn. A stick layer who is glaringly compared to his work mates would be stubborn indeed to ignore the need to improve. Peer pressure is a powerful force. Still, there are always those who refuse to change. Photo documentation over a reasonable period of time should be grounds for dismissal or reassignment. You cannot afford to have these people on the job. A final note: "Connect the stickers" on both sides of the unit and compare both sides. Stick layers often position their stickers in the far pocket and simply drop the end they hold. Look for wavy sticker alignment on the stick layer's side.

Figure 3.15 A single sticker laid on edge affects all of the lumber in its vicinity.

Avoid turning stickers up on edge. This causes serious distortions in the lumber (figure 3.15). If you notice this a lot, it may indicate that you are running the stacker too fast. Rushed stick layers cannot pay close attention to detail. Sometimes stick layers intentionally turn stickers up on edge or double stack stickers on one end to keep the lumber surface level. Look for one of two causes. First, are you sawing taper into your lumber? As layers mount up, one end rises faster than the other. Stackers do not operate well

Figure 3.16 Haphazard stacking resulting in uneven unit edges. The uneven edges affect airflow uniformity between layers when this side is next to the plenum chambers.

unless the lumber is level. Stick layers compensate for this by doubling up on the low end, usually without comment to the supervisor. Sometimes the stacker itself is not level or loses its level as it loads. I have seen this on stackers whose cables were replaced on one end only. The stretch difference between the two sets of cables increased as the load increased.

Figure 3.17 Good stacking technique, which helps maintain uniform airflow and seems to help the grade quality of the edge boards.

Ideally, we want a solid lumber layer across the unit width with even edges on the side for best drying. Wide variations in lumber target widths make solid layers with even edges difficult. Full-width layers often alternate with others lacking a board. There is variation in total width between full layers. Here we must compromise a bit to achieve our goals. Encourage stick layers to pull side boards to the outside to form an even edge (figures 3.16, 3.17 and 3.18). This improves airflow uniformity between layers. However, pulling boards to the side leaves a gap in that layer, forming a chimney.

Figure 3.18 Why are we missing a board here?

There is always a gap in 2 × 10s. These solitary edge boards are drier and prone to warp. Eliminate the chimney and reduce dryness by pulling back one board on the first layer, two on the second, one on the third, and so on as layers are stacked.

Be careful if you allow stickers to protrude on the outside. Mills operating tilt hoists in the planing mill will sustain serious sticker losses unless they turn protruding stickers to the outside. Exposed sticker ends simply shear off.

Care and Storage of Stickers

There are two common methods for transporting and storing stickers between the planing mill and stacker. Most transfer operations operate on a closed loop system, conveying stickers directly from the unstacker to the stacker. Many of these systems use a rake-off method of unstacking that is very gentle on the lumber and stickers. Sticker breakage is low. The stickers do not sit around on the ground and deteriorate. However, introducing new stickers to the system is often difficult unless provisions for this are designed into the original installation.

Most forklift operations use a tilt hoist method of unstacking. They bundle stickers or gather them into racks for transport back to the stacker. This is tough on stickers. Rising hoists shear stickers off on one end. As they slide onto the conveyor, weakened stickers break. As stickers are gathered into racks or strapped into bundles for forklift handling, still more breakage occurs. Storage takes its toll as bundles are heaped on one another. Stickers represent a significant investment for any mill. You must train personnel to protect that investment. Carefully stack stickers in racks or bundles. Store bundles off the ground and no more than one high (figure 3.20). For long-term storage, place them under dry shelter.

A good friend in North Carolina offers this wise advice to stacker supervisors: "Care enough about your stick laying to show that you are interested, today and everyday. Only then will you see lasting results" (figures 3.21 and 3.22).

Figure 3.19 If too few boards were sorted into the package, scatter the top layer to distribute the weight.

Figure 3.20 Proper use of a sticker rack.

Figure 3.21 Our goal—a perfectly stacked unit. The forklift driver is not too shabby either.

Figure 3.22 A well stacked and set up West Coast cart demonstrating their technique of placing stickers on 4-foot centers but adding additional stickers in the bottom layers for additional support. I tend to discourage the practice of stacking on 4-foot centers.

CHAPTER GUIDELINES FOR REDUCING VARIABILITY IN DRYING

1. Train your stick layers by showing them the effect poor stick laying has on lumber.
2. Determine the most economical sticker material for your operation by comparing service life to initial cost.
3. Use stickers $3/4$ inch to $7/8$ inch thick by $1\frac{1}{2}$ inches wide. Timbers may require thicker stickers to extend their service life.
4. Stickers must extend fully across the width of your units.
5. Remove thick, thin, crooked, and broken stickers from the system on a constant basis.
6. Align stickers one above the other in your units for complete support.
7. Lay stickers on 2-foot centers to align with truck supports.
8. If stickers begin appearing turned on edge or double-stacked, find out why.
9. As they stack each layer, stick layers should pull the edge boards toward them to form a square-edged package.
10. Finally, take care of your stickers between the unstacker and stacker to extend their life.

REFERENCES

1. Denig, Joseph, Stephen J. Hanover, and C. Arthur Hart. *Kiln Drying Southern Yellow Pine Lumber.* The North Carolina Agricultural Extension Service. Publication No. AG-402,1988.
2. Horton, Robert K., and Helmuth Resch. "Kiln-Stick Thickness Effect on Drying 2-inch Western Hemlock Lumber." *Forest Products Journal* 26, No. 3 (1976): 35–40.

Chapter Four

Kiln Setups and Loading Practices

Defeat doesn't finish a man—quit does.
A man is not finished when he is defeated.
He's finished when he quits.
Richard M. Nixon

Once lumber has been manufactured, sorted, and stacked, it is ready to load into the kiln. How well we set up lumber for loading has a significant impact on how well it dries. Our goal at this point is to dry the most lumber in the shortest time with the highest quality. This maximizes the efficiency of our operation and reduces costs.

Selecting Lumber from Stacked Green Inventory

Lumber going into the kilns must be as even aged as possible for highest quality drying (figure 4.1). Fresh-sawn lumber dries at a different rate from lumber that has been on the yard long enough to surface dry. Stained lumber dries differently still. Mixing lumber within the same charge results in wet and overdry units at the planing mill.

Inventory stacked lumber each morning and set up kilns from this list, drying the oldest lumber first. The ideal will be to process lumber into the kilns that is as green from the saw as possible, keeping pace with sawmill production. In general, consider as even aged all lumber manufactured within any seven-day period and dry this in the same charge. Lumber over a month old is considered even aged and is dried as partially air dried stock.

Design plenty of drying and boiler capacity into your sawmill system. Mills with marginal or inadequate kiln capacity leave no room for error. A

Figure 4.1 The lumber you place into each kiln charge must be as even aged as possible to avoid drying problems. Try to keep the ages within a single kiln charge within a seven-day production period.

well-balanced sawmill and drying operation is able to turn all sawmill production within six days. This leaves one day for routine maintenance and unexpected downtime. Without this cushion you can expect unforeseen problems to overwhelm you with green lumber.

We have already discussed several points to keep in mind when setting up your kilns. Let's review them here:

- Never mix thicknesses in the same kiln charge (figure 4.2).
- Mix no more than two widths within the same kiln charge. If you stack for forklifts, place the wider lumber on top. Dry 2 × 4s by themselves if possible.
- Never mix unit lengths on the same cart unless you fill the gap.

Keep your stacked green yard as small as possible to maintain control. This is easiest for forklift yards with computerized multizone kilns mixing widths in the same kiln charge. These mills often store green stacked lumber by length alone. If you must separate widths, then your yard must be much larger. Follow the guidelines in chapter 2 for setting up these storage yards.

Transfer operations need sufficient storage track for a complete charge of each width. This storage may include the green end kiln tracks when large volumes of a few widths are produced. These widths are loaded directly onto tracks to the next available kiln. Leave enough storage space for those odd

Figure 4.2 Keep stacked cart heights within two layers of each other and avoid mixing thicknesses in the same kiln charge.

units you drop out to get the right length combination for your kiln. Don't forget to put them into the next available kiln charge.

Find a system for tracking lumber from the stacker to the kilns that works for you. The options vary from spray painting item and stacking dates on the sides of units to attaching tags with this data. (Sample tags for this purpose are illustrated in chapter 13.) New, computerized systems read bar coded numbers to inventory lumber and set up kilns for optimum drying efficiency. They are also used to track lumber for the in-line moisture system in the planing mill. They are valuable for any operation handling large volumes of lumber.

One note on mixing different ages of lumber in a dry kiln. Many mills will observe an extension in drying time in the late summer/early fall and still see wet lumber in selected widths even though they are running the kilns longer. The extension in drying time is likely due to fungal growth on the surface of the lumber as it is being held for the kilns. During this season, if you are holding lumber (particularly the wider widths) for long periods to accumulate a complete charge of one item, you may have lumber going into the kilns with ages ranging from dead green to partially dry and molded. This is virtually impossible to dry to a uniform moisture content. Either mix widths to get the material into the kilns within a few days of sawing or build a small kiln to handle the drying of these small-volume items. This is an ideal situation for a single track kiln.

Kiln Carts

Let us first define some terms:

A kiln truck is the two-wheel assembly fabricated with two side channels. Two trucks are necessary to make one kiln cart. The bunk supports are the tubular cross beams that support the load. Many people, myself included, often refer to the complete assembly as a truck. The complete assembly is correctly termed a cart.

Good-quality kiln carts with supports under each sticker location are essential for high-grade drying. Four-foot-long carts built with two trucks and three bunk supports on 2-foot centers are the standard for the southern pine and other regions of the country. Many western mills seem to have no trouble with bunk supports on 4-foot centers, although I do not recommend them. In either case, the 4-foot carts are easily combined in different combinations to support all lumber lengths. This assures continuous lumber support and weight transfer to the ground for distortion-free lumber.

As mentioned, some mills place only two supports on a 4-foot truck. Their initial investment saving is often quickly wiped out through increased lumber degrade and shortened truck life. Bottom lumber layers tend to distort with a characteristic "kiln kink" typical of these operations unless the bunks are aligned perfectly with the stickers. The side rails of trucks bend as weighted ends force the channel to curve down under load. This distortion alters the total height of the cart, and new carts added to the system will be taller. As multiple cart heights mix together in the system, bottom layer degrade becomes severe. Once distorted in this manner, you must shim each cart individually to add a third support. New carts will still be higher. The best solution is total replacement of all kiln carts at one time. If this is not possible due to financial constraints, keep cart styles separate as you proceed with the conversion.

When buying kiln carts for the first time or as a total replacement in an existing system it is recommended that the carts be as rugged as possible. Standard trucks from the manufacturers are built using 4-inch side channels. This results in a truck that has no up or down, can be used either way, and is a throwback to earlier days when trucks were not normally welded into a complete cart. In the bad old days, trucks and bunks were physically moved from the dry end tracks and reassembled into carts on the green end. This is very uncommon today.

Today all trucks are assembled into the welded units we know as carts. If you have trucks with 4-inch side channels and attempt to put three bunk supports on the completed cart, there is no way to use them without hitting the truck wheels with the forks as they slide under the load. For this reason I always specify trucks with 6-inch side channels (figure 4.3). Although a lit-

Figure 4.3 The recommended kiln cart is 4 feet long, has 6-inch side channels on the truck to protect the wheels from forklift damage, and boxed beam bunk supports on 2-foot centers. This is a rugged truck that gives a long service life.

Figure 4.4 An old-style 4-foot kiln cart with only two supports on 4-inch side channel trucks. Note the bending in the truck channels from the lumber load.

tle bit of the wheel will often still protrude above the channel, it is not enough to catch the fork. These trucks will last a long time and are ideally suited for forklift operations. Transfer operations can continue to use 4-inch side channels, but they must use three bunk supports on the 4-foot truck to avoid side rail distortion (figure 4.4).

Use wheels fitted with grease fittings on the end of the axle and lubricate them at least once a week. Some operations grease them each time through the kiln. This is best done at the stacker in transfer operations, at

Figure 4.5 I-beams used as bunk supports in a forklift loaded system. When the forks are slid in they often catch the beam and the entire cart is picked up with the load. Notice the bottom layer distortion from insufficient lumber support.

the kilns in others. Previously, I recommended trying Vekton™ bushings rather than roller bearings for your truck wheels. New, they roll easily, and reportedly require less maintenance. Although the manufacturers claim you do not need to grease these bearings, you will get longer life from them by doing so. They are not maintenance free. Today, particularly in high-temperature applications, I recommend the tried and true roller bearings. They carry more load and can take a lot of abuse, but like anything else, they must be maintained on a regular basis. I have seen too many square wheels from seized nylon bushings and from ungreased roller bearings as well, for that matter.

The best bunk supports are made from square tube or from two pieces of channel welded to form a tube. The tubing should have a minimum 1/4 inch wall thickness. I-beam supports are easily bent in transfer operations and catch forks in forklift operations (figure 4.5). Reinforce the tubular bunks by welding on end caps. This is particularly useful for forklift operators who like to move loaded carts around by pushing on the ends of bunks with their forks. Weld bunk supports permanently in place to keep them in the right position. Welded trucks are easier to move about by forklift. Never use wooden bunk supports. They vary widely in dimension and must be handled and positioned each time they are used (figure 4.6). This degrades lumber.

The minimum recommended combinations of 4-foot kiln trucks for each length combination is shown in table 4.1.

Figure 4.6 Wooden bunk supports that are too few, of poor quality, and poorly positioned to support the load.

Table 4.1 COMBINATIONS OF 4-FOOT KILN CARTS AND LUMBER LENGTHS

	MINIMUM NUMBER OF CARTS FOR	
Lumber Unit Length	**Forklift Operations**	**Transfer Operations**
8′	2	2
10′	2	2
12′	3	3
14′	3	3
16′	3	4
18′	4	4
20′	4	5
22′	4	5
24′	5	6

I prefer to use the transfer system numbers for both operation methods. This assures maximum support with minimum slippage while moving loaded trucks about.

Avoid 6-foot trucks unless you stack 6-foot lumber. These trucks make odd combinations for setting up the various lengths going into the kilns. Purchase only a few to cover your 6-foot production. Be very careful with them, since some have insufficient support for the weight they are asked to carry. Either beef up the side rails or add extra wheels to avoid truck distortion. If you insist on using them, make sure bunk supports are on 2-foot centers to

Figure 4.7 An example of a single, full-kiln-length tram with moveable supports. These trams require close attention by the forklift driver to prevent degrade.

align with the stickers. I once visited a mill that was using 6-foot trucks with bunk supports on 3-foot centers. These trucks did not line up with anything. I occasionally find mills using 6-foot carts with bunks on 3-foot centers in response to forklift drivers who insist they cannot set up lumber with smaller spaces. They use the same argument to leave out crossouts between the loads. This is just an excuse to mask their not wanting to bother with taking the time to do it right. Besides, since when does the forklift operator dictate to the mill on questions of quality?

Use 8-foot trucks only at stud mills. Here you can place stickers and bunk supports on 32-inch centers.

I have seen many examples of a single, continuous cart, called a tram, for each kiln track (figure 4.7). One was for a kiln that was 84 feet long. Due to the large number of possible lumber length combinations to fill the truck, supports were unattached. In theory, this allowed the lift driver to position each support where needed. In practice, however, few lift drivers will do this consistently on a continuing basis. The grade losses from this truck were staggering. Some mills weld the truck supports in place. But be cautioned that it is difficult to place these supports in a configuration that accounts for all lumber length combinations. While I do not recommend this as an option, it is common in Canada and will work well as long as the person setting up the load takes the time to do his job right.

Whatever cart design you have, personnel must take the time to align bunk supports under the stickers. Illustrate this to your people by painting arrows pointing up at each truck support and arrows pointing down at each sticker. If the points of these arrows do not meet, you are losing lumber to

Figure 4.8 Mark a few units to visualize how loads are distributed to the carts. Arrows pointing down are the forces imposed by the stickers, arrows pointing up are the bunk supports. Where the arrows are unopposed the lumber will develop "kiln kink." This is also a tram.

"kiln kink" (figure 4.8). This shows up in the dry sheds and at the planing mill. Do not make the mistake of thinking only a few boards are lost by poor truck placement. Each edge board represents an entire layer, and several layers are usually affected.

Keep units aligned top to bottom in the kiln so they do not lean over. Units that are offset from the vertical upset airflows by varying the width of the side plenum chambers (figure 4.9).

Figure 4.9 Although this shot shows the area between the tracks, poor cart setups such as these alter the width of the plenum chambers and affect airflow uniformity top to bottom and along the kiln length. Try to make your setups as straight as possible.

Figure 4.10 Metal spacer to prevent cart shifting while moving carts.

Transfer operations stack their lumber directly onto kiln trucks at the stacker. They follow the same guidelines for use and placement of carts. Since the entire stacked cart may be moved great distances, these mills often have trouble with individual carts shifting out of position. This is particularly troublesome on units that have a 2-foot gap between neighboring carts. This always occurs in 10-foot, 14-foot, and 18-foot units. Mills that try to save on trucks see it in 16-foot and 20-foot units too. If the pull cable is attached to the truck or bunk support, trucks easily shift out of position. Avoid this by using the maximum number of trucks that will fit under the unit or by placing spacers between the truck gaps at the stacker. Some mills have the transfer operator cut and insert wooden spacers each cycle, while others fabricate steel spacers that travel with the trucks at all times (figure 4.10). A simple solution that often works is to position the space on the end farthest from the cable attachment. This makes it harder for cars to shift under the load.

Crossouts

Forklift operations load kiln carts at the kilns. In addition to positioning the kiln carts, personnel must also insert and align crossouts (also called bolsters) separating the bundles. A few transfer operations unload with forklifts and also insert crossouts at the stacker.

Crossouts should be good quality, straight, and unbroken. Since they serve to support the units above and rest on the units below, they must be uniform in size. Place one at each sticker location (figure 4.12). Failure to do this results in warped lumber in the top and bottom layers they contact

Figure 4.11 Everything you do wrong affects something else along the way in the system.

Figure 4.12 Insufficient crossouts to support the lumber.

Figure 4.13 The impact of one misplaced crossout can be remarkable.

(figure 4.13). Arguments from forklift drivers that there is no room to maneuver forks with all these crossouts are unfounded (figures 4.14 and 4.15). They simply do not want to handle the extra crossouts or make the effort. This costs the mill money. When you make the drawing that locates sticker guides at the stacker, also determine fork spacing. There is a spacing that works for all length combinations.

Figure 4.14 Make certain crossouts are installed at each sticker location. This photo illustrates one way to make certain the kiln is filled when handling different lengths.

Figure 4.15 Damage resulting from a combination of factors. Too few crossouts to carry the load, top layer of boards not scattered across the top, and stickers that are too narrow.

Figure 4.16 Note the impact one damaged crossout has on the lumber it supports. Crossouts must be culled just as stickers are.

The most common material for crossouts is dried and dressed softwood 4 × 4s. Although relatively inexpensive and readily available, they have a short service life when exposed to day-to-day kiln conditions. Softwood 2 × 4s are also often used, although they tend to fall down if stood on edge. It is better to lay two on top of each other or nail two together. The main problem with both these crossouts is their tendency to break. Damaged crossouts that are not removed from the system degrade the lumber they contact (figure 4.16).

An alternative crossout material is square aluminum tube (figure 4.17). Large and stiff enough to support lumber without bending, these crossouts are light and easy to handle. Although their initial cost is high, they pay back in extended service life and improved grade. One mill I know had been using green gum 4 × 4s as crossouts, with an average life of three months. Based on the purchase price for green gum and aluminum tube, we determined that if the aluminum would last one year it would pay back the investment. These crossouts have now been in continuous service for over 15 years with no loss of quality or usefulness. Unless accidentally damaged by equipment (they do not support forklifts driven over them), they have a very long service life.

Keep in mind that these crossouts never leave the kiln area. Kiln personnel place and remove them for each charge, keeping a rack handy for their storage. Dry 2 × 4s are used in the dry sheds. The 2 × 4 crossouts see only one trip through the sheds, since they are dressed along with regular production in the planing mill.

Figure 4.17 Tubular aluminum crossout. Photo courtesy of International Paper.

Four inches by four inches is the maximum recommended size for crossouts. The smaller, the better, since a significant amount of air is lost through these spaces during the drying process. Some mills have found they can reduce crossouts to $1^3/4$ inches by changing to wide, thin, high-strength steel forks. Due to the width of the fork you may have to eliminate two crossouts to have enough room, and this degrades a little lumber. A little diesel oil on the forks helps them slide in easier. Covering crossouts speeds drying by increasing the average kiln air velocity. It also narrows moisture distributions by eliminating the very dry boards surrounding the crossout space.

Table 4.2 PERCENTAGE OF TOTAL AIR VOLUME LOST THROUGH THE CROSSOUTS

Number of Crossouts	Typical Lost Air Volume
1	11.0%
2	16.8%
3	26.6%
4	39.7%

Over the years I have run many tests to determine the effect of closing these crossouts. In one test we took air velocity readings with the crossouts open and found the average velocity was 975 fpm with a velocity range of 310 fpm. After pushing the lumber back out and covering the four crossouts with 1 × 4s, we reloaded the kiln and tested airflows again. Average velocities had increased to 1,118 fpm for a 15% increase in airflow (figures 4.18 and 4.19). The range reduced to 250 fpm. Keep in mind that you will never recover all of the air volume you calculate you are losing through the crossouts. By closing them you increase the static pressure that the fans must work against. This reduces the total volume of air the fans are able to generate. Still, if you have multiple crossouts, this is a simple way to increase the efficiency of your drying system.

Blocking this space is a benefit (Table 4.2), but it can be time consuming. For the test, we covered the crossouts with green 1 × 4s, but this is not a realistic procedure on a day-to-day basis. Other mills take short lengths of 2 × 4s and place them in the opening. Another mill stuffs short sections of fire hose into the space or staples it over the opening. Another uses foam rubber blocks stuffed into the opening. All work fine on crossouts you can reach from the ground. Once you must tote and climb a ladder to install blocks, however, the work and time involved outweigh the benefit in many operators' minds. Mills with only one crossout are lucky. This space is easily reached from the ground.

You may be able to minimize or eliminate crossouts altogether. There are several methods in use. We have already discussed going to thin forks and using 2 × 4s on their sides to reduce the opening. There are air bladders available that are inserted into the crossout opening and inflated to raise the

Figure 4.18 Lumber crossouts covered for an air velocity comparison check.

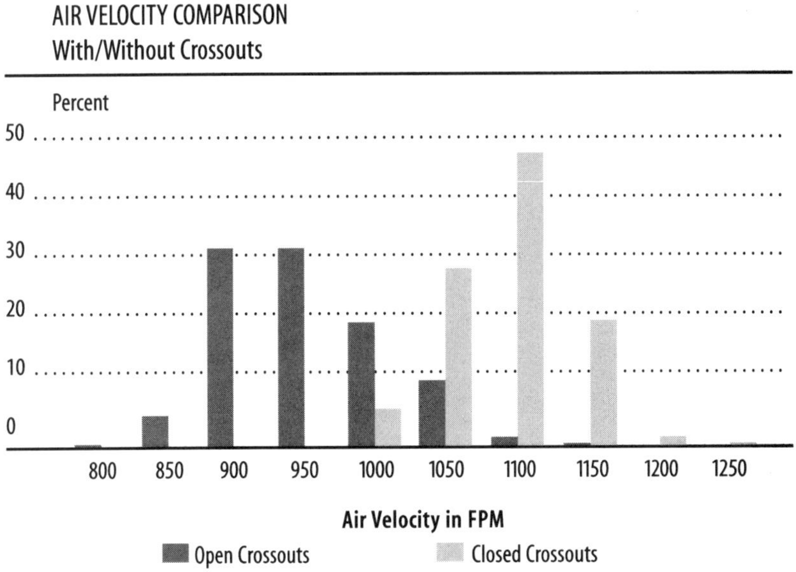

Figure 4.19 Shift in air velocity by covering the crossouts in figure 4.18.

lumber off the crossouts while the crossouts are replaced with stickers (figure 4.20). The bladders are then deflated and removed. This requires a dedicated worker whose sole job is to perform this task. This has become a burden to many mills. Most have abandoned it.

Figure 4.20 Using an air bag to remove crossouts and replace them with stickers.

The easiest option I have heard of along these lines is being done at a southern mill, although I have not personally witnessed it in operation. Reportedly the lumber is either set onto the kiln carts adjacent to heavy vertical uprights, or the carts are moved to the uprights after loading. At any rate, the uprights are positioned between the forklift and the load in such a way that once the crossouts are removed and replaced with stickers the lift simply backs out, leaving the lumber on the stickers. To remove the lumber after drying, the load is simply speared with oiled forks.

If you have a single crossout, there is a closure method that is simple and easy to install. At this mill, aluminum strips 1/8 inch × 4 inches × 10 feet hang horizontally on chains against the wall. Wherever the chain attaches to the strip, a clip made from heavy strapping is attached. This is the strapping normally used on chain rail cars. These clips slip into the sticker spaces above and below the crossout space, holding the aluminum strips in place. The nice thing about this method is the lack of clutter on the floor, since the chains keep the strips against the wall and at a convenient height. They are quickly removed and reinserted when changing the kiln.

Never leave crossouts on top of the top units within the kiln. Overhead baffles rest on them and leave a large avenue for air loss.

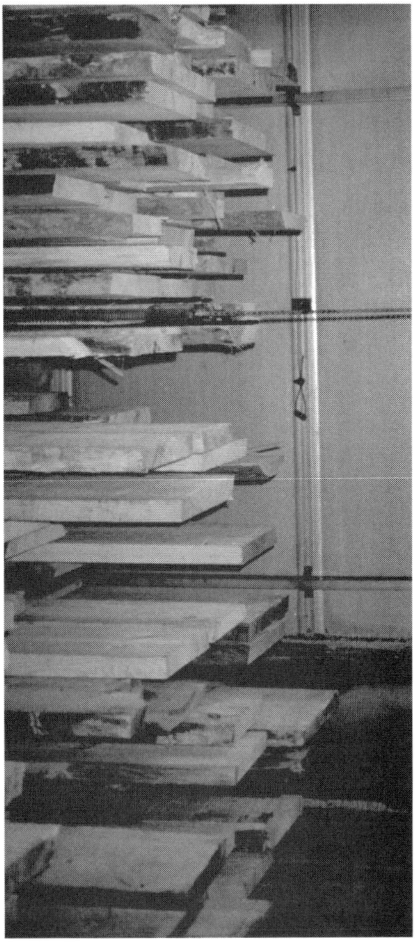

Figure 4.21 Short loading your dry kiln hurts you in two ways. It represents a lost drying opportunity that you may have to make up later, and it diminishes and disrupts the airflow in the charge that is short.

Important Loading Considerations

Always fill kilns to capacity. Plan ahead. Once you start loading, it is usually too late. This is the difference between managing an operation and just running it. Your kilns were designed to hold a maximum amount of lumber—anything less is a lost drying opportunity. Short loading a 68-foot double track kiln costs you 1.5% capacity for each 2 feet of short track. If every charge is short 2 feet, then you must dry an extra five charges a year to make up what you lose. If the charge is so short that your end baffles do not contact the lumber, airflows become unbalanced and uneven drying results (figure 4.21).

When forklift loading, always set units to the very edge of the kiln trucks on the side next to the floor

Figure 4.22 Set lumber units to the edge of the kiln bunks so that the floor baffles will contact the lumber and close off any gaps.

Figure 4.23 Uneven load heights make it difficult for the top baffles to close the gap. Plan out your kiln loading instead of simply letting it happen.

baffles. This ensures a tight fit with the baffles and eliminates the short-circuiting of air under the lumber (figure 4.22).

Make certain that the total height of the loaded trucks is uniform from one truck to the next (figure 4.23). Changes in height of one or two layers are not serious, but avoid drastic changes that result in gaps at the top where overhead baffles contact. This is a potential source of air loss. Baffles must contact the lumber to do their job.

Units of adjoining trucks must butt tightly end to end to avoid air loss gaps. This is difficult at mills with poor lumber length control, resulting in the kiln operator trimming ends at the kiln to minimize the gap or just to close the kiln doors. Before trimming, however, determine that the lumber is indeed over length and not just poorly even-ended at the stacker. Trimming these results in trim loss. If poor even-ending is the problem, these boards will usually push back into the units while loading. You will still have to trim boards to close the doors unless they are easily driven with a sledgehammer. However, if the exposed ends shatter, you might as well trim them. This is an opportunity to get the people in the sawmill or at the stacker involved in correcting the problems they cause. Once you identify the source of the problem, make sure that those responsible see and understand the impact at the kilns.

Kiln Loading and Pulling Methods

Just as there are two primary methods of kiln setups, there are two methods for loading, depending on whether the system is transfer or forklift loaded. The method used at any particular mill is strongly influenced by the personal preference of the mill designer. There are advantages and disadvantages to both.

Transfer Systems

Transfer operations are labor and equipment intensive, requiring specialized equipment (transfer cars) and reasonably skilled personnel to do a good job. The primary advantage is the uniformity of stacking, which reduces the impact of crossouts and poor setups on airflow. This advantage disappears when kiln personnel fail to fill kilns completely and leave gaps within the kiln (figures 4.24, 4.25).

Use extreme caution when working around these operations, since cables used to pull lumber have a tendency to slip off or break. Another hazard is walking between units on the same track. I recall one instance in which a man was crushed between units as the kiln was being loaded. He was not discovered until after the load was dried because he was not a part of the reg-

Figure 4.24 Poor loading practice. Note the crossouts left on the top of the unit, which leaves an additional gap for air loss.

Figure 4.25 Poor load planning with poor follow-through.

Figure 4.26 When you set two units side by side on a cart, leave sufficient space between them for airflow to redirect as it passes through.

Figure 4.27 Transfer operator moving lumber onto the track for loading into the kiln.

ular workforce. Everyone who works with these systems quickly becomes attuned to the dangers (figure 4.27). Give moving cables a wide berth.

Transfer cars have a tendency to become very temperamental in their operation, either through poor design or deteriorating maintenance. As an existing transfer system ages, the ends of tracks next to the transfer pits wear and are often patched. Concrete in the vicinity of these repairs chips out, offering little support. Truck wrecks become more frequent as slop wears into the system from the uneven forces exerted by moving large volumes of lumber by cable.

Here are a few guidelines for designing a reduced maintenance transfer system:

1. Rim all exposed edges of the transfer pits with heavy 3/8-inch-thick 3 × 3 angle iron. This stabilizes the edges and gives something for the cables to wear against.
2. Bury the rail in concrete up to the ball of the rail, and slope track slightly in the direction of lumber movement through the kilns. Never build valleys between tracks for drainage. You will never be able to get a forklift in to correct truck wrecks. A flat, even surface ensures forklift maneuverability and is a lot easier to work around.
3. Use the heaviest cable that personnel can comfortably work with. A cable that is too light breaks easily and must be replaced often. Use the cable with the least stretch to reduce the tendency for lumber to overshoot the kiln. This creates additional work.
4. Install pneumatic or hydraulic cylinders at each corner of the transfer cars (figure 4.28). Once the car is aligned with the tracks, these cylinders extend and stabilize the car in position, reducing the likelihood of wrecks.

Figure 4.28 One of four hydraulic cylinders mounted on the corners of a transfer car. These cylinders stabilize the car as lumber is moved onto and off the tracks.

5. Pull the dry lumber out of the kiln before pulling the green in. A few mills pull both at the same time and have higher than necessary transfer car and cable repair costs.

What is the best method to attach cables to the loaded kiln cart? All methods are a compromise, but you have two basic choices. Either you attach to the cart or you attach to the lumber.

The most common method of attachment to the cart involves simply hooking the cable on the end of the bunk support. This is not recommended since cables easily slip off the end, and bunk supports bend. This method also imposes a twisting motion to the cart, which contributes to wrecks and damaged wheels and bearings. A better method is to lay a cable between the tracks with a hook on one end to hook the car and a loop on the other to attach to the winch cable. This requires two complete cable sets per track. Before the carts to load green lumber are placed on the tracks, a cable is laid between them. With the tracks loaded, this cable is used to pull the lumber into the kiln and is left in the kiln during the cycle. A second cable is laid between the green tracks and lumber loaded onto the tracks as before. When the dry lumber has been pulled from the kiln, that cable is moved to the green tracks and the process repeats. Individual units of dry lumber are easily moved onto the dry end transfer car with cables on the side. Any system that attaches cables to the trucks must make provisions to prevent carts shifting underneath the lumber. This may be as simple as moving the space between carts to the end opposite cable attachment. Here, you must place longer lengths on the end next to the door, since 10-foot lumber almost always shifts its carts. If you cannot control the sequence of lengths going onto the tracks, then you must insert spacers between the carts to prevent shifting.

Attaching cables directly to the lumber prevents shifting carts, but usually results in mechanical damage to lumber and/or stickers. There are several techniques. Cables may loop around the corners about waist high and hook onto an end bunk support, pulling up on the truck. This breaks corner boards and stickers. Other mills drive bars or wedges into the end sticker spaces. Then a cable is run between or beside the track and attached. In this case the ends of boards are damaged. Some mills intentionally notch the ends of lumber for cable attachment. The best system for this I have seen uses a heavy flat plate specially designed to insert in the sticker space. The bearing plate is wide to spread the force and reduce board damage. A cable is run under the truck and attaches to the plate. This works pretty well unless the device slips out. Fishing for a new cable under these trucks is not easy. It is doubtful that the people who designed most transfer systems ever ran one.

Forklift Systems

You will find two common methods for loading kilns by forklift. The first is set up similar to a transfer operation, in that one end of the kiln is designated the green end and the other end is the dry end. The dry end usually has a cooling shed. These systems must move empty carts from the dry end back to the green end for loading. Occasionally you will see a transfer system on the green end with a forklift removing dry lumber and returning trucks to the stacker. These systems either push in and out by forklift alone

Figure 4.29 Cart damage resulting from pushing on end of bunk supports with forks.

Figure 4.30 Kiln pulling techniques that minimize lumber and equipment damage.

or use cables in conjunction with the forklift. Both systems are cumbersome and require precise maneuvering by the driver. They also put wear and tear on the carts (figure 4.29).

The loading system I prefer has no designated green or dry end of the kiln (figures 4.32, 4.33, and 4.34). A double track kiln loads green lumber on the track at one end of the kiln and on the other track at the other end. This

Figure 4.31 Kiln pulling techniques that damage lumber or equipment.

permits access to both sides of either track for loading. To push in, a bar is placed into notches built into the end trucks, which face each other when the kiln doors are opened. Or you can use 4 × 4s or 6 × 6s simply laid on the tracks. The forklift operator then backs up to the first line of green lumber and pushes (figure 4.33). As he pushes the green lumber in, the dry lumber is pushed out (figure 4.34). The spacer(s) is moved to the next line, and the

Figure 4.32 Pulling method for kilns that do not have a green or dry end. A piece of shafting is placed in brackets built on the ends of the carts. Some mills simply lay two 8-foot 4 x 4s or 6 x 6s on each track. Photo courtesy International Paper.

Figure 4.33 Well-stacked and setup lumber minimizes kiln degrade problems.

process repeats. This is very fast and efficient, allowing you to turn the dry kiln in 30 minutes or less. Truck damage is minimal, since carts spend their entire lives on the track. There are two things to keep in mind, though. First, there can be no spaces between carts within 20 feet of the end where the bar is positioned. Otherwise, trucks shift out of position as the load is pushed. Second, it is often necessary for the lift to pick up a bundle of green lumber before pushing to increase traction. This is particularly important in the winter.

Figure 4.34 The forklift backs up to the green lumber and begins pushing. Photo courtesy International Paper.

Figure 4.35 Once the dry lumber is out on that track, the shafting is moved to the second track and the process is repeated. This method eliminates lumber and cart damage. Photo courtesy International Paper.

CHAPTER GUIDELINES FOR REDUCING VARIABILITY IN DRYING

1. Select lumber for each charge that is as even-aged as possible. Dry your oldest lumber first.
2. Always fill kilns to capacity.
3. Use 4-foot kiln carts with three supports spaced on 2-foot centers.
4. Place sufficient kiln carts under each unit of lumber in such a way that every sticker has support. Make sure the supports align with the stickers.
5. If you use crossouts to set up your carts, place one at each cart location. Discard any damaged crossouts that affect uniform lumber support.
6. Cover crossout openings to reduce air loss through these openings.
7. Never leave crossouts on top of units going into the kilns.
8. Bring lumber to the outside edge of the cart bunk supports on the floor baffle side to permit good floor baffle contact.
9. Keep the total height of loaded carts uniform for good overhead baffle contact.
10. Butt the ends of units tightly together to reduce air loss between them.
11. Align units top to bottom to improve airflow uniformity top to bottom and along the kiln length.
12. Adopt a kiln pulling method that reduces damage to lumber (figures 4.30 and 4.31).

Chapter Five

Final Loading Considerations

Mistakes are a fact of life.
It is the response to error that counts.
Nikki Giovanni

Now that lumber is pushed into the kiln, we become more interested in certain design features of our drying equipment. Are we getting the full potential from our baffles? Do we lose too much heat through the doors or elsewhere? The answers to these questions have a significant impact on the quality and speed of our drying.

Everyone who runs a kiln sees, and should inspect, the inside of the kiln each time it is loaded. However, most operators position their baffles and close the doors, never asking if their equipment actually performs its intended purpose. Have you ever entered your loaded kiln with the fans blowing and verified that the equipment is doing its job?

Kiln operators sometimes take a lot on faith, assuming that the kiln manufacturers have designed the right equipment to do the job. Never assume. Most of the engineers who design kilns have certainly never run them. Many have never seen them in operation. The next time you prepare to start your kiln, take a moment to look around. Load the kiln as usual, positioning baffles and closing the doors. Lock out the heat system, turn the fans on, and close the vents. Now, walk down both plenum chambers and look. Take an air velocity meter if you have one. Do the baffles stay in position? Are they performing their intended purpose? How much air are you losing through gaps in the baffles or lumber? If you have crossouts, how much air is lost there? Do you feel cold air coming in around the doors? Do you see any light around the doors and vents? Is everything as tight as it could be? Make a note of the items you find amiss, and repeat your inspection with the fans blowing in the opposite direction. Correcting these problems goes a long way toward improving your drying operation.

The Baffle System

Baffles are hinged metal doors that are positioned against the lumber to close gaps between the lumber and floor, fan deck, and kiln doors. Their sole purpose is to direct all available air through the lumber (figure 5.1). Baffles that do not contact the lumber or that move out of position during operation are not doing their intended job. Inspect your baffles each time you load the kiln. You have to handle them anyway, so pay attention to how well they are working.

Baffles must be durable and capable of taking abuse. Baffles constructed of 16-gauge steel or heavier with edges broken for stiffness or otherwise reinforced are good. They tend to be a little heavier to handle, but stand up well in service. Avoid aluminum baffles unless you have severe corrosion problems. They simply do not stand up to wear. Aluminum overhead and floor baffles also require additional weight or restraint to keep them in place. Again, this may not be apparent unless you inspect your kiln loaded for running and with the fans on.

Figure 5.1 Properly designed and positioned overhead baffle with the end baffle folded flat against the lumber.

Overhead Baffles

Overhead baffles must fully contact the top of the lumber and extend to, but not over, the edge. These baffles smooth the transition from horizontal airflow out of the fans to the vertically aligned plenum chambers.

Baffles that extend over the edge restrict airflow through the top layers. Indeed, I have seen overhead baffles that literally envelope the top three to six layers, resulting in an engineered wet spot (figure 5.2). This step is often taken to compensate for lumber units that are chronically overdry at the top. The real problem stems from poor control over heat and airflow distributions. Correct this at its source by multizoning to control heat variation top to bottom in the kiln if possible. Lowering your dry bulb set points by 5 to 10 degrees may also help. During your inspection, you may be surprised to discover that overhead baffles "sail" or float above the lumber. This reduces average air velocity, so weight them down.

If the overhead baffles contact the lumber a few boards back from the edge, you will find the exposed edge boards overdry and warped (figure 5.3). The top edges of units also present a nonuniform profile to airflow, which creates turbulence. More serious, as the lumber dries it shrinks, and the overhead baffles slip between the top board gaps. When these baffles are raised, boards often dislodge and fall. They are then lost from the system due to breakage or excessive warp and counted as downfall. The safety hazard to the operator is real. I know: I was once knocked to my knees as a dislodged board glanced off my shoulder during an inspection. You learn very early to stay close to the outside walls as you raise baffles. A few mills construct elab-

Figure 5.2 An overhead baffle that envelopes the top layers. These top layers lay flatter at the end of the drying cycle because they are wet.

orate mechanisms designed to protect the operator from falling boards. The real problem stems from improper baffle placement and the poor drying conditions that result. Correct the problem and enjoy the benefits in better drying and reduced degrade.

Remember, if the height of the lumber in your kiln varies over two layers between adjacent units, then you are losing too much air over the top. I have entered a few kilns to discover overhead baffles dangling well above the top layers of lumber. This is a sorting, stacking, and ultimately a management problem.

Figure 5.3 An overhead baffle that does not come out to the edge of the load exposes the top edge boards to conditions that degrade it.

Figure 5.4 Never allow the overhead baffle to rest on the end baffle. This gap will grow larger as the lumber dries and shrinks. Always lower your overhead baffles first.

Figure 5.5 The design of this overhead baffle is fine. Poor loading has reduced its effectiveness.

Never leave crossouts on the top where they will hold the overhead baffles off the lumber. Some mills do this to stabilize the load and reduce dislodged lumber. Others do not recognize the harm in this practice, assuming the overdried top layer is lost anyway. This is an open avenue for air loss.

Repair damaged baffles as soon as possible after discovery. Large holes introduce variation to the drying system and affect drying uniformity. The most common source of damage results from failure to raise baffles when changing out a charge. Many mills compensate for this potential oversight by tying all overhead baffles to a single winch and pulley system, raising all baffles at once. I do not recommend these systems, since gradual, uneven wear results in baffles with varying clearances. Undetected cable failures or

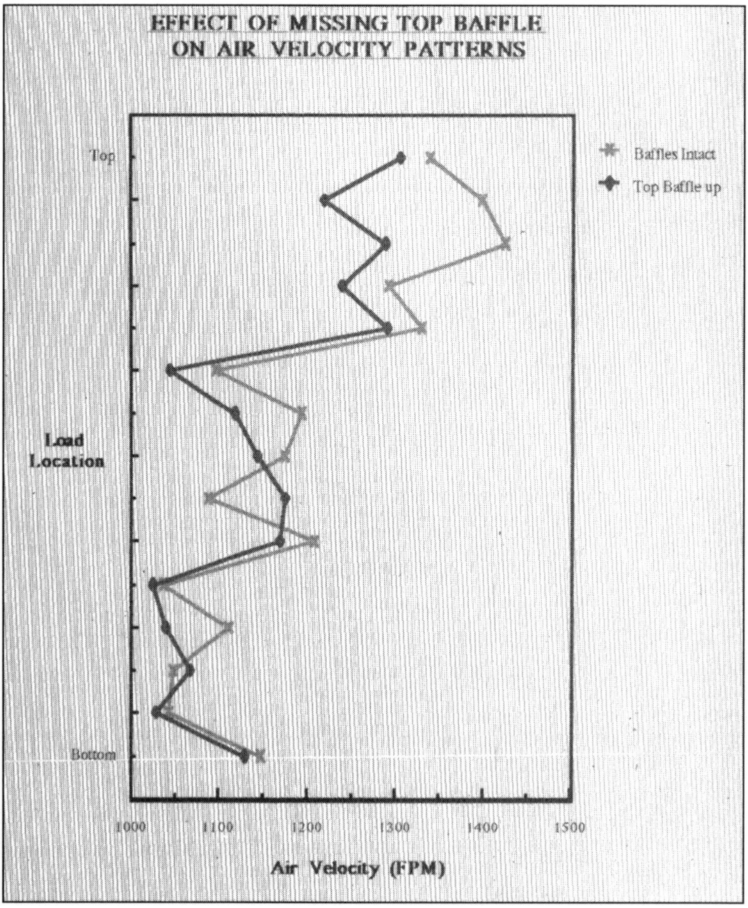

Figure 5.6 Variation in air velocity top to bottom in a kiln with and without the overhead baffle. Note the significant impact on the top one-third of the load.

an extra-tall unit will still rake off top layers and destroy baffles. The only difference is that you can blame it on that "#$%&! cable" rather than poor memory or inattention. Pull each baffle individually to ensure proper clearance. You do not need to wipe out too many baffles before you remember they are there, particularly if you are responsible for repairs.

How much do missing top baffles affect your drying? Figure 5.6 illustrates the change in air velocity off the second track with the baffles down compared to with the top baffle up.

Floor Baffles

Floor baffles close the gap between the floor and the bottom course of lumber. There are two main types in general use: poured-in-place concrete and hinged metal baffles. I prefer metal baffles because they completely close this gap and do a better job of reducing air loss under the lumber. They also exhibit better airflow characteristics towards the bottom of the load.

Concrete baffles seem to be a carryover from early conventional kiln designs. These kilns were built with line shaft fan systems and dried lumber with low temperature and air velocities. Stationary overhead baffles and no end baffles were standard. If you still have one of these kilns, it is possible to dramatically increase their airflow by installing hinged overhead and end baffles. At one mill running a conventional triple track kiln we increased average air velocities from 175 fpm to 345 fpm by simply installing hinged overhead and end baffles. They also had concrete floor baffles. But we had to live with them.

The main problem with concrete baffles is that they cannot be moved. In order to provide clearance for the passage of trucks these baffles are installed with a 3-inch or more gap between them and the truck (figure 5.7). Combined with sloppy lumber setups, I have seen this gap stretch to 9 inches or more. A large volume of air moves through this space and affects drying efficiency. To find out how much, measure the air velocity coming through the gap and compare it to that coming through the lumber. Close the gap by using metal baffles that rest on the concrete and slide against the lumber. Attach handles to aid in placement and reinforce their edges with angle iron or bend (break) them for stiffness. A properly designed concrete floor baffle has its top surface even with the top of the truck, and these metal baffles will contact the bottom board. The idea here is to close the gap without restricting airflow through the lumber (figure 5.8). You may have to make modifications from the flat design to achieve this goal in your kiln. If you have a computerized multizone kiln controlling on exiting air you *must* close this gap.

Figure 5.7 Typical gap left by a concrete floor baffle.

Figure 5.8 Be careful that you do not create additional problems for yourself when you close the concrete floor baffle gap.

Figure 5.9 A properly designed floor baffle should just contact the bottom-most board at the bunk support. Note how the end baffle has been augmented with belting to close the space at the bottom.

Figure 5.10 Poor floor baffle design. Top edge of baffle leaves a gap under the load. The units were not placed on the carts properly either. These units must come to the edge of the bunk on the baffle side.

Figure 5.11 Floor baffle held off from the load by a protruding sticker. When the wind gets behind these baffles, they will often fall over. When loading the kiln, break off these stickers so the baffle will lay flat against the load.

Figure 5.12 Repair damaged floor baffles when discovered.

Figure 5.13 Metal floor baffle, which must be slid against the load. If one person must move this baffle in position, it reduces the likelihood that it will be positioned correctly.

Hinged metal floor baffles give the tightest fit. The main objection I hear is that, like overhead baffles, they are subject to damage and wear. Properly designed and cared for, however, they last a long time. The benefits of their use far outweigh their repair cost. The floor baffle design should have the baffle contacting the bottom board with enough slope on it to prevent its being flipped back by the airflow (figure 5.9). Look for laid-down baffles when you pull your kiln and determine why those baffles fell. Common causes are boards or stickers overlapping the edge of the truck and warped baffles. When baffles are held at a near vertical angle they fall easily when the air catches them. To correct this, position lumber just to the edge of the bunk support. Replace warped baffles. Some mills elect to install restraints to keep them from flipping over under any circumstance. They require extra handling when pulling the kiln. If you have one of the new, super-high-velocity kilns, you may have no choice but to use them. Whatever works for you is the correct approach. Just make sure your floor baffles are doing their intended job.

Figure 5.14 Hinged wooden floor baffle.

I have seen floor baffles that varied from 2 × 12 lumber (figure 5.14), to building siding, to corrugated water pipe simply rolled into place and held by a brick. With the exception of the 2 × 12s, which is a common practice on the West Coast, none of these baffles were designed into the kiln but were added as an afterthought. Get rid of them and install floor baffles specifi-

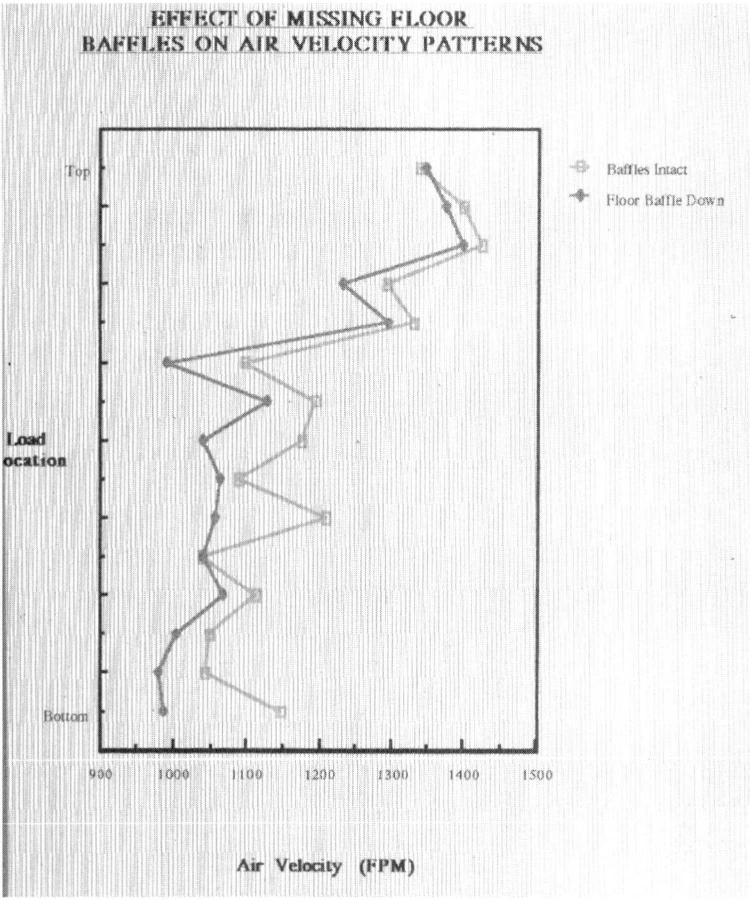

Figure 5.15 Variation in air velocity, top to bottom, with and without floor baffles. As expected, the impact is felt nearest the bottom.

cally designed for your kiln. The payback is faster, higher quality drying (figure 5.15).

Vertical End Baffles

End baffles prevent air from flowing around the ends of lumber next to the doors. Unless they securely close this gap, they are useless (figures 5.16 and 5.17). Most end baffles are bi-fold, although single-fold and tri-fold baffles are also found.

Single-fold baffles are found at those mills that have made a firm commitment to always fill their kilns to capacity. They benefit from maximum kiln throughput and improved drying because these baffles do not block air-

Figure 5.16 A single-fold end baffle at a mill that has not quite made the commitment to fully load its kilns.

flow through the last 2 feet of lumber like bi-folds and tri-folds will. They are easy to hold in position, but they leave little room for error.

Bi-fold doors are the most common and accommodate variations in load length up to 4 feet. In fully loaded kilns, most operators fold the extra 2 feet of baffle flat against the lumber. If they determine lumber dryness

Figure 5.17 Although this end baffle is properly positioned against the load, notice how the placement of the crossout has reduced its effectiveness.

by metering the ends next to the doors, they must compensate for the potentially higher moisture content of the end. If the baffle's angle iron hinge lies against the lumber, the shrinking unit often binds it. This dislodges overhead boards, which may strike the kiln operator as the baffle is opened. The alternative is to angle the outer baffle edge against the last

Figure 5.18 A properly positioned end baffle with solid, positive restraint.

sticker position and secure it there. This avoids blocking the last 2 feet of lumber but opens a large hole at the top and bottom for air loss. Bi-fold end baffles are particularly useful for mills that stack mixed length packages. By placing the mixed length ends next to the door, the end baffles help control the overdrying that can occur in the ends of these units by restricting airflow through them.

Manufacturers maintain that tri-fold doors accommodate variations in load length up to 8 feet. I question this in high-velocity, high-temperature kilns. These baffles expose a significant surface area to the airflow, and I have never seen one with sufficient restraint to keep it in place. Once the fans are turned on, they move about at will.

Inspection of your baffles with the fans blowing often reveals glaring problems with positioning and restraint. When placed into position they must contact the outside edge of the lumber, ideally resting against the last sticker. Never fold the baffle against the end of the unit. This looks like you are doing something, but once the fans are on they flatten against the inside of the door and lose their function (figure 5.19).

The key to the effectiveness of any of these end baffles is adequate restraint (figures 5.17, 5.18 and 5.19). The baffles must remain in position, tight against the lumber, and block all air movement in their direction. The restraint must be firm, but flexible enough to adapt to changing loads. The spring restraints supplied by the kiln manufacturers are little more than glorified screen door springs. New ones have a nasty habit of snapping back in

Figure 5.19 End baffle improperly positioned against the end of the unit. With the air blowing toward the viewer, this baffle flattens against the inside of the door. The restraint serves no purpose.

Figure 5.20 Note how the end baffle attachment point in this kiln leaves a large gap between the door and the baffle, rendering the end baffle useless. This is a prime example of what might be designed by an inexperienced kiln engineer.

your face. They never seem long enough to reach the spot where you need to hook them. Fortunately, or unfortunately, they lose their spring tension quickly and stretch out. Then...well, you get the idea.

There is a better way. This idea is the brainchild of Jim Brown, supervisor at a local mill. Jim simply uses a piece of 3/16-inch or 1/4-inch chain about 12 feet long for each baffle. He attaches it to the baffle halfway along its length to distribute the load. He hooks the other end onto a bracket inside the center door mullion or wraps it around an extended board, sticker, or crossout after the baffle is pulled tight into position. He uses a small chain binder to pull up the slack, but this is not necessary (figure 5.21). The method is safe, simple, inexpensive, and—best of all—permanent.

Figure 5.21 Chain-style end baffle restraints are secure and permanent. Their effectiveness is increased by securing the chain near the vertical midpoint of the end baffle. The loose end can be hooked to another chain or simply wrapped around a protruding board.

Figure 5.22 illustrates the impact on airflows from improperly restrained end baffles.

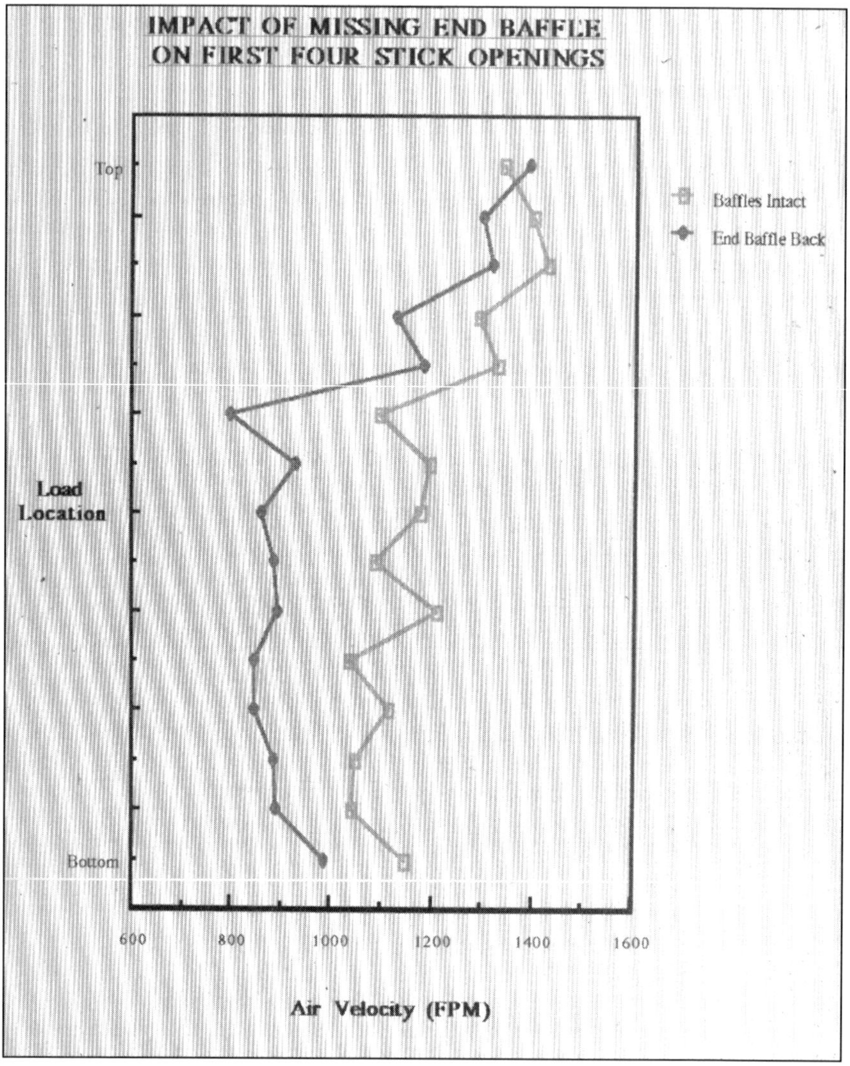

Figure 5.22 Variation in air velocity, top to bottom, in a kiln with and without an end baffle. The velocities are an average of the first four sticker openings.

Figure 5.23 A poorly designed end baffle with improvised restraints. Notice how the lumber is visible through the hinge. Too much lumber is covered by the baffle.

Figure 5.24 Replace end baffles when damage is discovered.

Figure 5.25 These packages are ideally placed next to the end baffles to help block air loss through the load.

Top Loading

This is the practice of placing weights on the tops of your stacked units to help restrain the top layers from warping during drying. This technology originated in Australia and New Zealand and has been shown to be effective in reducing top-layer warp. I have seen top load ranging from 1/8 inch and thicker flat steel plate, to walkway grating, and all the way up to steel reinforced concrete slabs from 4 to 12 inches thick (figure 5.26). To gain maximum effectiveness it is recommended that these loads remain in place until the lumber has been allowed to cool.

My personal opinion is that top-layer warp stems from a lack of control over the drying conditions in that area of the kiln. These upper layers typically see higher average air velocities transferring more heat and contributing to overdrying induced warp. In multizone steam kilns this warp is best reduced simply by using schedule set points that allow center coils to run in temperature control. Some control systems allow the operator to bias the center coils to place more or less heat into different areas. Direct-fired kilns present a different problem. The main heat distribution duct in these kilns is mounted above the lumber stacks. Temperatures in this duct typically approach the 450 degree range, which radiates from the duct onto the exposed top layers. Simply insulating the underside of this duct should go a long way towards minimizing top-layer warp in these kilns. Blocking air delivery from the upper quarter or so of the downcomers has also been beneficial. These options will be discussed in greater detail in the section on direct-fired kilns.

Some guidelines for the use of top-layer weights:

- Use loads in the 20 to 40 pounds per square foot range.
- Leave the loads on as long as possible after drying to allow the lumber to cool.
- For the heavier loads, stick laying, crossout placement, and placement of units on carts must be perfect. Any deviation compounds warp and "kiln kink."

There is an internally mounted top-layer restraint on the market, which is constructed of 1/8-inch plate and is raised and lowered with air cylinders. The design typically includes overhead baffles. Although a bit pricey, these restraints do improve the flatness of the top layers, even though you cannot allow the load to cool with them in place. If you are considering their purchase, I suggest that you first investigate controlling this warp through your schedules before making an investment of this magnitude.

Figure 5.26 An example of very large top-load restraints.

Loading Package-Style Kilns

In my discussions of kiln drying I admittedly give short shrift to this style of dry kiln. These kilns were originally intended for use by hardwood mills to dry lumber that had been air dried before kiln drying. Owing to the large volume of lumber they typically hold and the limitations on the fan and heat systems that result from their basic design, these kilns are generally not recommended for drying green-from-the-saw softwood lumber. Their primary advantage is their low initial purchase price. You can purchase a high-volume kiln for a remarkably low price.

From a drying quality standpoint, they leave a lot to be desired. Air velocities and heat delivery are typically low and are hampered by extremely long air passes (typically 16 to 20 feet) before the air stream can be reheated. When viewed from the load end, the slowest drying area of the kiln is the bottom center, the fastest is the upper corners. The difference in moisture content between these two areas can be quite remarkable and determines how fast the kiln can be run.

When loading these kilns the principles are the same as for track kilns. Load without gaps and with the overhead and end baffles contacting the load at the front and rear (figure 5.27). Air must be forced through the lumber with no open avenue around it. Many of these kilns have stationary baffles at the rear top and sides. These must overlap the load area (keep in mind that lumber will shrink as it dries, so you want to load your heights to take advantage of this fact). When loading the first line of lumber, I like to justify it to the left and use a long vertical baffle to close the gap on the right. The second line is justified to the right and the third to the left again, and so on. This will break up gaps that always remain when loading these kilns. You should never be able to walk front to rear in the kiln through these gaps. The last line of lumber is justified to one side, and the end baffles are positioned to close the gap. There should be hinged overhead baffles at the front of the kiln as well.

Spacing between the individual lines of lumber is crucial to good airflow. I recommend that you paint guidelines on the floor for the forklift driver. When painting these lines work in a 4- to 6-inch gap between each line of lumber.

Since the forklift drives into these kilns for loading and unloading, the kiln walls typically sustain a lot of damage. I remember a line of four package kilns at a mill some years ago where you could walk between them from one end to the other and not have to go outside in the rain. Sheathing the interior side walls with steel plate will cut down on this damage and give you something to load against.

Figure 5.27 Loading a package kiln follows the same principles as for track kilns. You want no open avenues for air loss around the lumber.

Figure 5.28 This mill has added a middle baffle to its package kiln to cover the random-length ends.

The width of the front and rear plenum chambers is very important to obtaining good airflow in these, as it is in any kiln. The wider the better, but no narrower than the combined top to bottom sticker, and crossout thickness. Since these kilns are typically very tall (upwards of 21-foot door heights are not uncommon), this can result in a very large plenum area. These plenums must be balanced in width, front and rear. For those of you who stack 4-foot-wide packages, you must resist the temptation to add that extra line of lumber at the front. The additional capacity is not worth the disruption it causes your airflow.

Kiln Doors

Kiln doors act as a movable, insulated, solid barrier to separate the internal kiln drying conditions from the outside. A remarkable number of mills seem to lose touch with that fact. Door panels are knocked out, frames bent, hold-ins broken. Repairs take months, if they are done at all. Keep in mind that the only reason we really bother to construct a kiln at all is to gain control over the conditions that regulate rapid, quality drying. If we lose that, then we defeat our purpose. Holes and loss of seals introduce cold, unconditioned air to the kiln environment—air that is neither sensed nor considered with conventional drying controls. Repair any damage to doors as soon as possible to maintain control (figure 5.29).

Anyone who runs a kiln for even a short while soon wonders about all the steam that escapes around the seals of even a new door. A conventional tadpole gasket has a very short service life due to the mechanical damage it sustains during normal handling. Alternative materials, like fire hose, are more resistant to mechanical damage but break down quickly in the kiln. Older kilns are locked into some sort of gasket material for life and should follow the manufacturer's recommendations. Manufacturers must be challenged to develop sealing materials with built-in mechanical protection. Until then, fiberglass rope wrapped in fiberglass fabric seems to work best in high-temperature kilns. The silicone gaskets will hold up to conditions as well, but all are subject to mechanical damage as doors are closed.

New kilns, however, have another option. Well-designed kilns can avoid using gaskets altogether in favor of a metal-to-metal seal. This gives the best seal possible, but is highly dependent on the quality of design and fabrication to achieve matched surfaces. The first few charges seem to steam excessively around the doors as surfaces conform to each other. Over time, you can rely on a good seal without worry about damage and replacement. There will be leakage, but it will be far below conventional gasket methods. Never attempt this on an older kiln; there is simply too much distortion to be successful.

To increase the effectiveness of a metal-to-metal seal, run a heavy bead of silicone sealant around the perimeter of the door facing where the door rests and cover it with waxed paper. Close the door. Once the silicone has cured you can remove the waxed paper. This will fill most surface irregularities.

Figure 5.29 Poor door closure resulting from sprung doors.

With the doors closed and the kiln empty, inspect the inside of the doors. If you can see light, then you have identified a leak source. If it is at the top, check to see that the door is on the top hooks. These hooks slope down toward the kiln to use the door's weight to help hold it tight against the jamb. If it is on the hooks, but there is still light, check the bottom hold-ins. These occasionally fill with dirt and support the door off the top hooks. They must allow free movement up and down. Finally, check the side mullions and their hold-ins. They must be snug against the side pins. These should be adjusted only after you have determined that the door is fully down in the top hooks.

The spot where the tracks run under the doors has always been a problem. The common response is simply to notch the bottom of the door. This weakens the door and leaves an opening for heat loss. In addition, high-velocity, moisture-laden vapors stream out of this opening, eroding the exposed track. Complete deterioration often occurs in 6 to 8 years. Years ago one manufacturer simply slotted the rail at the threshold to allow the angle on the bottom of the door to slide into it and avoided cutting notches in the bottom of the door. This is an excellent idea that I have not noticed in years. Another alternative is lift-out rail sections under the doors (figure 5.30). Pioneered by Frank Cook on the West Coast, this innovation is occasionally seen in the South. Until 10 years ago I had not seen one I would recommend. Now I know the secret to success is in the foundation. Both the kiln and loading aprons must be rigidly aligned at the doors, or the two will move out of alignment as the kiln settles. This alignment is absolutely criti-

Figure 5.30 Lift-out rail section at bottom of door. Since the door bottom need not be cut, the door seal and strength is improved. In this position, the rail section acts as a stop to prevent lumber from being pushed into the kiln door.

Figure 5.31 Door sill deterioration covered with old belting.

cal. Without notching, the integrity of the doors is maintained, and tracks do not erode. An added benefit is that they act as load stops, which prevent the forklift driver from accidentally pushing lumber into the doors while loading the next charge.

Erosion at the bottoms of doors affects not only tracks but the concrete thresholds as well. In new kiln installations, and when replacing sills, always rim the threshold in angle iron or preferably aluminum. This results in a more uniform seal and, as such, reduces erosion to the point where it is not an issue. As a temporary measure, stuff gaping holes with burlap bags or cover them with belting, cut to shape, to close obvious gaps (figures 5.31 and 5.32).

Door carriers should allow a single man to lift and move the kiln doors easily. This is a function of the design skill of the manufacturer. Being of slight stature, I quickly learned which manufacturer's doors I could move without strain, and which I could not. There are a few I cannot move, period. Keep door carriers greased on a monthly basis (insist—no, demand—that new door carriers be supplied with grease fittings at all pivot and rolling points). In the full lift position, the bottom of the door must clear the track. If it does not, then something is wrong with the design. It is up to the manufacturer to correct it. I have seen kiln operators strain to pull the door out at the bottom as they try to move the door to the side. Aside from the obvious toll on the operator's body, the carrier wheels twist on their axles, causing premature carrier failure.

Doors are known to slip off their carriers and fall for any number of reasons. This is quite surprising to the operator, and he is in danger of serious

Figure 5.32 Severe door sill deterioration, which must be rebuilt.

injury. OSHA requires that all doors be equipped with restraining brackets at the top to prevent such an accident. These brackets tend to become bent out of position over time, particularly if you have recently replaced your doors or have done major repair work. Keep their purpose in mind.

It is because of poor design and operator frustration that recent years have seen the introduction of barn-style doors (figure 5.33). These doors are split into two halves and hinged at the sides. Due to their size, their hinges

Figure 5.33 An example of a barn-style door.

must be well designed not only to carry the weight, but also to swing easily without distorting the door or the jamb to which it is attached. Alignment is critical to satisfactory service. Their major drawback is that they must clear the tracks at the bottom as they swing out. This leaves a large gap, which must be closed in some manner. This usually involves some sort of gasket or belting arrangement with unknown long-term sealing problems. Any barn-door design must adequately address all the seal, alignment, clearance, and maintenance issues without compromising the primary purpose of a kiln door: the separation of the outside from the internal kiln environments. After 10 years of building these doors, most of the manufacturers seem to have gotten their designs down pat.

Access Doors

Ground-level operator access or "escape" doors are another OSHA requirement (figure 5.34). I have never known of an instance where they were used in an emergency situation, yet it is comforting to know they are there. I can assure you that, faced with the prospect of being trapped in a running kiln, I would fabricate my own door in very short order.

These doors are very handy for entering a kiln to check airflows, change wet bulb socks, or troubleshoot problems. Most operators open them to help cool the kiln before entering. Having an access door at each corner of the kiln is handy. Aside from the safety aspect, they allow you to pass quickly from one end to the other when changing out your kiln. These doors are best located on the end corners or side wall rather than on the larger doors. Those located on the main doors force you to unlatch the baffles before entering and are a nuisance. They become particularly awkward when trying to check air velocity because you must relatch the baffles from the inside to get an accurate reading.

A few manufacturers place another access door on the gable end wall above the main doors. This permits ready access to the fan deck for inspection, maintenance, and repairs. If you have ever had to squeeze through a fan deck trap door or climb between the overhead coils on the side plenums, you will appreciate this feature. The obvious advantage to the mill is that, having eliminated the ordeal of getting to the fan deck area, maintenance is kept at a higher level.

Figure 5.34 Ground level access doors are mandatory and handy when placed at each kiln corner.

CHAPTER GUIDELINES FOR REDUCING VARIABILITY IN DRYING

1. Overhead baffles must contact the top of the lumber along the entire kiln length. They must not overhang or extend back from the edge.
2. Floor baffles must prevent all air loss under the kiln trucks.
3. End baffles must remain in position when the fans are turned on and close the gap between the lumber end and the doors.
4. Replace damaged baffles and restraining devices immediately upon discovery.
5. Maintain kiln doors and their seals to reduce air infiltration into and out of the kiln.

Chapter Six

Dry Lumber Handling

In order to succeed we must first believe that we can.

Michael Korda

Once lumber is dried and pushed from the kiln, we must allow sufficient time for it to cool and equalize. Paragraph 161.4 of the SPIB 1977 rule book states that "Kiln dried lumber shall be adequately cooled after removal from kilns before dressing and final moisture content testing." How long does it take before lumber is "adequately cooled"? That really depends on how bad the planing mill needs the lumber.

The longer you can let your lumber sit under shelter before dressing, the fewer problems you will have with it later. When lumber is first removed from the kiln, each piece has what is known as a **moisture gradient**. Have you ever watched the dial on a resistance moisture meter as you drove the needles slowly into a just dried board? At first the moisture reads low, but it gradually increases as the needles penetrate deeper. This difference in moisture content between the surface and core is the board's moisture gradient. In high-temperature lumber, this gradient can be very wide, with the surface being bone dry to a core approaching 20% or higher.

As the lumber sits in the cooling shed, the moisture migrates from the high-moisture content (MC) core to the low-MC shell. Allowed to sit long enough, the moisture evens out. If you drive your needles into the board now, you should find little difference between shell and core. The lumber has equalized within itself, the moisture of the wet core migrating to the dry surface.

But we have a planing mill foreman who just has to have the lumber that is in the kiln *now*. He cannot wait. He paces the floor moaning and groaning, worrying that the kiln will not finish before he runs out of dry lumber. He glances over toward the kiln, sees you opening the doors for your first look, and rushes over. "Is it dry? Is it dry?" he asks impatiently. After you complete your checks, you reluctantly conclude that he can have it. (You actually found a few units that weren't quite ready. Under less pressure, you might have run the kiln a bit longer to be sure.) He is jubilant, rushing to the lift or transfer operator to impress upon him the importance of getting that lumber to the planer "RIGHT NOW!!!" You, my friend, have been had.

A week or so later the customer to whom this lumber was shipped calls and complains about excess moisture. The plant manager calls you and the planing mill foreman into his office to find out what happened. He turns to the planing foreman, who explains, "He"—meaning you—"said it was dry. We didn't bother to check it because you know how hard it is to check hot lumber. Besides, we would have had to shut down the planing mill because we were running out of lumber."

The manager turns to you. Slowly you begin. "Well, I did check the lumber in the kiln and from my experience it seemed to be dry," you begin. "We had a couple of units that were a little high, but I've never had a problem with them before. I just don't understand what happened."

What did happen? Think about the purpose of a planer. It is designed to plane off the surface of the lumber to give you a uniform, smooth dimension. Since it removes the surface, it is removing the driest material from the board. There is nothing left to offset the high moisture content of the core. When the lumber does equalize, it will do so at a higher moisture content than it would have, given more time. While we all realize that the ultimate responsibility for lumber quality rests with the grader who allowed it to pass, the blame will eventually lie at your feet.

Let your lumber sit, under shelter, for a minimum of 24 hours. I prefer two to four days. Some mills have determined that seven days is optimum. There has to be a reasonable balance. Most transfer car systems are designed with extensive track and shed space behind the kiln for this. Forklift operations occasionally have a cooling shed over the dry end where lumber is allowed to set overnight. If not, then lumber must be removed to the dry sheds as soon as it is pushed out of the kiln. Hot lumber must not sit in the rain. The dry surface will soak up a lot of water very quickly. If you run out of shed space, place protective film on top of units as you move them into storage. This helps them shed water.

In the dry sheds, set up your supports as you would for stacked green lumber as described in chapter 2. Place support sills so that the lumber does not span over 4 feet. The lumber must rest on the sills so that the stickers align with them (figures 6.1). Crossouts must also align with the stickers and the sill positions to eliminate distortion. Unlike green lumber, dry lumber is not flex-

Figure 6.1 Proper positioning of bunk supports under the stickers, however, there are not enough of them to reduce lumber damage. This lumber needs four supports.

Figure 6.2 Improperly placed foundation sills for dry lumber. Degrade and trim loss is assured in addition to nicked planer knives.

Figure 6.3 Dry lumber should be held under dry storage until it can be sent to the planer for dressing.

ible and breaks if not evenly supported. Keep dry lumber off the ground where it might pick up moisture, rocks, or grit (figures 6.2, 6.3, 6.4 and 6.5).

Rough finish lumber held in storage for long periods may well pick up atmospheric moisture during damp periods or humid seasons. Check this

Figure 6.4 Sloppy dry lumber handling. This has become a safety issue.

lumber before it is sent to the planer. You may find you have to put it back into the kiln briefly to drop the moisture below the required 12% MC. If this becomes a chronic problem, you may be forced to store this lumber in an enclosed, and perhaps heated, storage building.

Figure 6.5 Keep your rough dry storage organized for easy retrieval of the next production run. Lumber should set a minimum of 24 hours to cool and equalize before being sent to the planer.

Figure 6.6 Poor storage of dressed lumber.

Figure 6.7 Take care of your dressed lumber to avoid moisture pickup, particularly in the higher grades.

CHAPTER GUIDELINES FOR AVOIDING PROBLEMS WITH DRY LUMBER STORAGE

1. Grade and pave the area under the sheds. Fill potholes. Avoid storing dry lumber outside without rain covers. Keep the area clean and debris free.
2. Use sills made from rot-resistant wood, concrete, or steel, spacing them to line up with stickers and crossouts to eliminate sagging. Support this lumber on 4-foot centers at a minimum.
3. Place crossouts at each sticker location or at least at the sticker where each foundation sill is positioned.
4. Check rough finish lumber for moisture pickup prior to sending it to the planer.
5. Let your lumber sit a minimum of 24 hours to equalize before sending it to the planer. The longer the better.

Chapter Seven

Fan Systems and Airflow

The beautiful thing about learning is nobody can take it from you.

B.B. King

Airflow is arguably the most important factor affecting the performance of any dry kiln. For many softwood species, the faster we can pass air over the lumber surface, the faster it will dry. High airflow also improves moisture uniformity. In training sessions I spend more time on this topic than on any other simply because here is where the major improvements and potential gains reside. After a brief description of the various fan system options available, the primary focus of this chapter will be to help you obtain peak performance from your existing system.

The purpose of the fan system is to deliver heat to, and remove moisture from, the lumber surface. Not only do we want maximum airflow, we also want uniform airflow, which is important for even drying. One of the first things I discovered as I studied dry kilns was that areas of high velocity within the dry kiln resulted in low moisture contents for the lumber in those areas. Likewise, low-velocity areas resulted in high moisture contents. While we want the highest possible average air velocity through our lumber, we also want to reduce its variability. This will result in the shortest possible drying times and narrowest moisture distributions we can attain with our equipment.

Fan systems for kilns are divided into two distinct categories: line shaft and cross shaft. Line shaft fan systems mount multiple fans on a single shaft, which extends the length of the kiln, and use a single, high-horsepower motor mounted at one end. Cross drive fan systems use multiple drives and motors with the shafts perpendicular to the kiln length. There are two cross shaft options. Direct-drive fans mount the fan directly onto the motor shaft, with the motor and fan assembly mounted entirely within the kiln environment. Shaft-driven fans mount the fan on a separate shaft, which is belted to a motor mounted exterior to the kiln. There are two variations on this as well. The first extends the shaft out the side wall, where the motor is belted to the exposed shaft end. The second uses a shorter shaft belted to a ridge-mounted motor.

Line Shaft Fan Systems

Line shaft fan systems are common in older kilns (figure 7.1). The first kilns I ever ran were line shaft, all 11 of them. The oldest had been built in 1951, the year I was born. Over the past decade I cannot recall having seen any new kilns built with these drives, although I do know they are still being manufactured, particularly for longer drying West Coast species. Even out west, though, the trend is towards cross shaft systems, which offer higher air flows and greater efficiency. Also, many older kilns originally built with line shafts are being converted to cross shafts to improve their performance.

With all of the fans mounted on a single shaft and rotating in the same direction, the same fan style cannot be used along the entire shaft length. Each fan must alternate between left- and right-handed. With the shaft rotating, alternating fans blow in opposite directions. It is through the use of baffling between and around the fans that airflow is directed through the load. Shaft rotation is reversed to reverse airflow.

NOTE: To tell the difference between a right- and left-handed fan, stand in front of the fan with the shaft near your body centerline. While facing the fan, note which of your hands slips between the blades in the direction they are pitched. If your right hand slips easiest through the blades, then it is a right-handed fan and vice versa. You cannot change a left-handed fan into a right-handed fan simply by turning it over. The fan is made for one hand or the other.

Due to the long length of the shaft and the number of fans mounted on it, shaft speeds are kept low (figure 7.2). As such, the maximum air speed through the load rarely exceeds 600 fpm. Usually it is much lower. Since the efficiency of most fans increases with their rotational speed, we can possibly squeeze a little more out of them by pushing the RPM. However, I strongly caution you to consult with the kiln manufacturer, fan manufacturer, professional engineer, or all three to see if your particular system can handle it. This option is not for the timid or faint of heart.

On older line shaft systems, increasing RPM is the extent of what we can do to gain more airflow from the system, since the fans in these kilns are

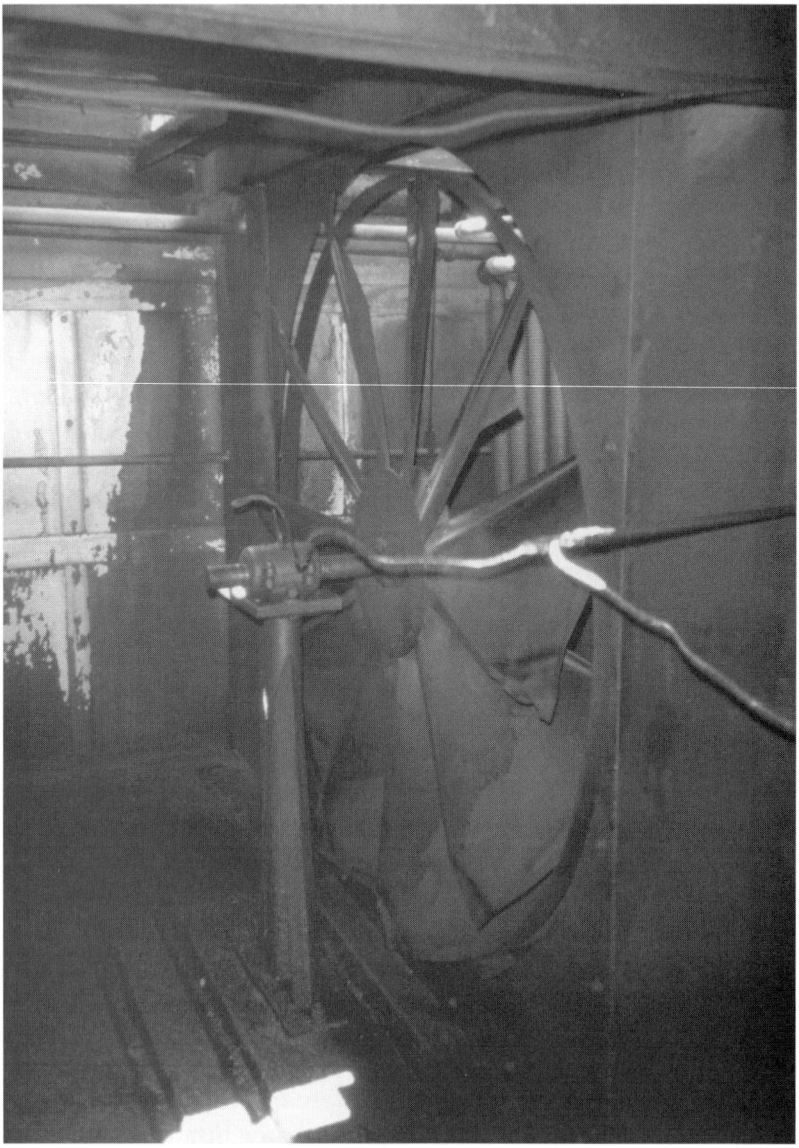

Figure 7.1 Line shaft fan system equipped with the older-style disc fans.

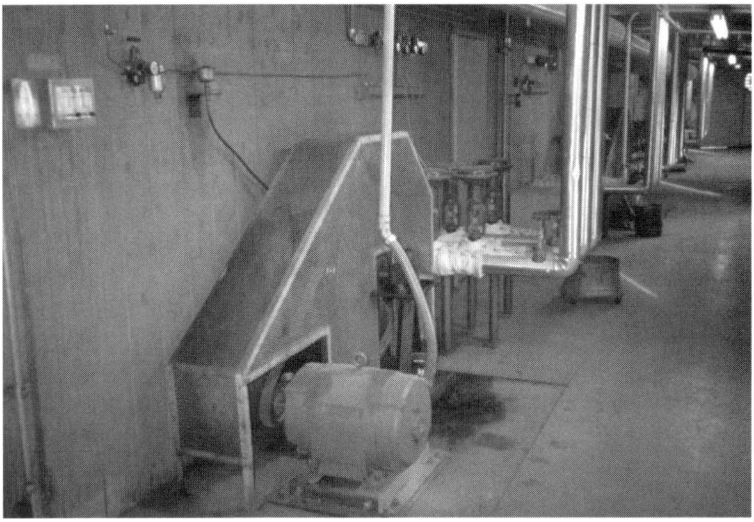

Figure 7.2 Drive for a line shaft fan system.

commonly fixed pitch. Newer kilns may be equipped with adjustable pitch fans, which give us a little more to play with. If we know we are running the maximum RPM, then we can begin to add pitch to the fans to increase the loading on the motor. Check the rated amps for the motor on its nameplate, and compare it to the actual amperage draw with the kiln loaded for running. If you are not running at maximum amperage and can increase fan pitch, then follow the directions for doing so found later in this chapter. Remember: a small change in fan pitch will have a tremendous impact on amp draw.

Finally, if you want to spend some money for a higher horsepower motor, you can probably get more out of the existing fan system. First, talk to all those folks I mentioned above to see if this is an advisable thing to do. If so, install the motor (you may have to enlarge the service), check the amps, and repitch the fans to maximum amp draw. That is the limit of what I would recommend for line shaft fan systems.

Later on we will talk about such things as tip clearances, shroud design, and so on. Owing to the low RPM involved here, unless shrouds are missing in whole or in part, or the fans are running outside of the shrouds, there is probably not a whole lot left to be done that would give you any added benefit. I have added rolled baffles to turn the airflow between fan bulkheads, but I cannot honestly say we were able to accomplish anything by it. Sure looks good, though.

Cross Shaft Fan Systems

Direct-Drive Fans

The next generation in fan systems were those with cross shafts. First came the direct-drive motor systems. Kiln manufacturers found that by mounting multiple motors and fans within the kiln they could increase air delivery over what was possible with conventional line shaft systems (figure 7.3).

The motors designed for in-kiln duty are commonly open drip proof, mounted in a motor housing one size larger for the horsepower. The windings are triple dipped and the leads are extra long and utilize high-temperature wire. This design allows wire connections to be made up on the roof, outside the kiln environment. Properly installed and maintained, these motors work well and have long service life at temperatures up to around 195°F. They are ideally suited for conventional schedules. I have run occasional accelerated schedules to 210°F but suspect that this contributes to premature motor failure. It is generally agreed that these motor systems are not suited to high-temperature applications.

Motors mounted within the kiln are exposed to high heat and humidity conditions. At one mill running two high-temperature kilns with seven fans each, we logged 18 to 21 motor failures per year. We were able to justify conversion to externally mounted motors based on rewinding, repair, and downtime costs, even though we purchased new motors for the system. It is dismaying to watch multiple motor failures occur on startup, knowing the work that is involved to change them.

The primary source of failure is not so much the heat but the high humidity of the kiln. Most failures occur immediately on startup. The reason is simple. Between charges or after a long idle period, the motor windings become relatively cool. As the kiln is loaded, moisture begins to condense on the motor housing and windings, as well as on the interior of the kiln. This is a particular problem in kilns that share common walls. Often the insulation and vapor barrier qualities of the common walls are gone and moisture vapor passes freely between the kilns. When the power is turned on, the motor shorts to ground. Prior to a startup on the above-mentioned kiln, we once tested the continuity of each leg of the motors in the kiln. All were shorted to ground. In a panic, we turned on the heat to evaporate the moisture we found dripping from the motors. We lost no motors on startup but concluded that something definitely had to be done.

A few manufacturers add circuitry to the fan system to place a 24-volt DC current on the windings while the motors are idle. This keeps them warm and prevents condensation. This precaution is highly recommended for all internal direct-drive fans.

Figure 7.3 Direct-drive fan equipped with a four-bladed prop. Six-bladed props are recommended on 10 hp and higher motors. Note the older-style shroud design.

Changing motors in these kilns challenges the stamina of the best millwright. Unless your kiln has access doors mounted on the gable ends, motors must be hoisted through a manhole. They must be manhandled and dragged along the fan deck (often corrugated), over and under steam pipes, sprinklers, and structural supports. Then they must be lifted into place. Just thinking about it makes me tired. If you have several kilns sharing common walls, this becomes very interesting—particularly if an adjacent kiln is running. Steam billows into the fan deck area, condensing on everything. Visibility—forget comfort—suddenly becomes important. To do anything, you must either shut down the offending kiln or at least set its fans to blow away from your work area. An overhead rail system for moving the motors through the kiln is recommended and much appreciated by the maintenance men replacing these motors.

We will cover the options for improving performance of these fans at the end of this chapter since these options are shared with the remaining drive systems.

Shaft-Driven Fans

Shaft-driven fans seem like the answer to every kiln operator's prayers. In general, they are. But you must be very careful when considering the different designs available. Some create their own problems with respect to uniform and maximum airflow.

The best performing systems, in terms of airflow, center the fans directly along the centerline of the kiln. All line shaft and direct-drive cross shaft systems I have seen align their fans in this manner. Why this makes a difference has to do with the fan manufacturer's recommendations on how far away from a fan you should begin turning the air stream. They recommend three fan diameters' distance. Most kilns must compromise on this a bit, but as long as the fans are centered, fan performance is similar regardless of fan direction. Performance differences increase as fans move away from the centerline. At least one kiln manufacturer, in the interest of economy, moves its fans off-center to reduce shaft and bearing costs, placing the fans somewhere over the middle of one of the tracks. Resist this temptation to save a few bucks. Airflow measurements in these kilns show significant variations in airflow patterns between the two fan directions. This affects final moisture distributions. Always insist on fans mounted centerline of the kiln.

External motors are mounted either at the eave or on the ridge of the kiln. Each is belted to the fan shaft. Ridge-mounted, also called through-the-roof fan systems, are a recent development; the jury is finally in on long-term performance.

"Through-the-Roof" or Ridge-Mounted Cross Shaft Fan Drives

Through-the-roof fans are mounted on a short shaft (36 inches or so), centerline or slightly off-center in the kiln. There are only two bearings. The sheave end of this shaft is boxed in and enclosed from the shaft to and through the kiln roof, where the motors are mounted. Drive belts pass through this protective enclosure and connect the motor to the fan shaft. They are less costly because of the short shaft (36 inches versus ±18 feet) and reduced bearings (2 versus 4 or 5) (figures 7.4 and 7.5).

However, there are several concerns with these drives…

- First, the belt enclosure is very large (18 inches to 21 inches wide) and positioned close to the fans. This blocks airflow through a portion of the fans' arc. As each blade passes this obstruction it unloads, causing it to flex and vibrate slightly. At normal RPM each blade flexes 850 to 925 times per minute, potentially leading to metal fatigue. Although I have heard no reports of excessive blade failures from metal fatigue, we do know that this design violates one of the primary rules for designing an efficient fan system by placing a large obstruction directly in front of the fan. More about that later.
- It is very difficult to tell if a fan has thrown its belt or has other problems. Installing an ammeter on each motor helps and allows you to compare current draw between all the fans in the system. Ammeters also allow

Figure 7.4 Through-the-roof style cross shaft fan system, external view. Note the handy catwalk and overhead rail system for easy motor maintenance.

Figure 7.5 Internal view of a through-the-roof fan drive. Note the short shaft and minimum of bearings, which lowers the initial cost for this drive. This is false economy. The sheave shrouding robs fan performance and forces you to pay more over the long haul.

you to detect slippage due to gradual belt deterioration. Monitoring and recording amp readings on a regular basis must be part of your preventive maintenance program with these systems. Most mills rely on visual inspection to discover fan losses. This is complicated by the shrouding around the belts, which makes visual inspection difficult to accomplish on a regular basis. It is recommended that if belts are worn on one fan, you should replace them all at the same time. This avoids the constant uncertainty one-at-a-time replacement brings and results in better overall system performance. This repair should be scheduled rather than imposed by sudden failure. Ridge-mounted fan systems require a high level of operator attention for proper maintenance.

- Once a belt failure is detected, changing the belt is difficult. You must shut your kiln down and allow it to cool or wait for the beginning of the next charge to make repairs. This costs you more downtime. These shrouds are not vapor tight, so expect more frequent belt failures due to the action of heat, moisture, and wood chemicals. These belts must be tensioned more to overcome their thermal expansion, and this further hastens belt failure.
- I hear increasing reports of high bearing failure rates with these drives, although I have not been able to pin down a cause, whether it is maintenance or the application. The additional belt tension may be a factor.
- Finally, expect to pay 15 to 20% more to obtain the same airflow in a kiln compared to an eave-mounted drive. Put another way, expect to get 15 to 20% less airflow from this drive system when compared with an eave-mounted drive running the same horsepower. This is a direct result of belt shrouding discussed in item 1.

So, you thought you saved money on your initial investment. Well, as they say, you can pay me now or you can pay me later.

Eave-Mounted Cross Shaft Fan Drives

In my humble opinion, the ideal external motor fan system mounts the motor at the eave of the kiln. This external cross shaft design has seen the longest service history. A long shaft with multiple bearings transfers power to the fan, which should be mounted central to the kiln. These systems are more expensive than the through-the-roof design due to the extra shaft length and additional bearings. Properly designed, they have the best service and performance record. Belt replacement requires only a short interruption of service. Also, we get more air per belted horsepower, since there are no obstructions in front of the fan as in the through-the-roof systems (figures 7.6 and 7.7).

Figure 7.6 Cross shaft fan with an eave-mounted motor drive. Obstructions near the fan are kept to a minimum. Note how the manufacturer has rotated the supporting beam to cut down on air turbulence.

The shaft is mounted on a large tubular or I-beam steel base extending from the fan to the outside edge of the kiln. This support must be heavy enough to maintain bearing alignment as well as transfer its load to the kiln structural members without distortion. Tube supports must control not only the vibration about the shaft axis but also the fan thrust. Heavier shafts increase stability and reduce the number of required bearings. Split bearings are preferred, eliminating the need to strip the shaft to replace them when they do fail. Early designs placed a special bearing, known as a thrust bearing, next to the fan. Its purpose is to absorb the thrust of the fan and maintain its position. This bearing should be located next to the sheaves to control belt alignment. This is more critical to fan performance.

Fans and Motor Horsepower

The horsepower for line shaft systems is dependent on shaft length, style and number of fans, shaft diameter, rotational speed, and desired static pressure. If you are running fixed-pitch "disk" fans, you can likely trade up to variable-pitch fans and improve fan performance. Horsepower recommendations for these drives are beyond the scope of this book since horsepower is specific to the individual kiln and must be engineered to the application.

Figure 7.7 Eave-mounted drive equipped with air bags to maintain belt tension.

For 10 hp cross shaft fan systems and up I recommend no less than 72-inch six-bladed props. Lower horsepower motors must use four-bladed props, also 72 inches in diameter. Although the fan curves indicate that you should not be able to get any more air out of 10 hp motors with six blades than four, in actual practice I find that the improved performance justifies their cost.

NOTE: Never purchase fans that have fixed pitch for any dry kiln. Only accept adjustable-pitch props. Fixed-pitch props give you no latitude to maximize fan performance by increasing fan pitch. They are a dead end for your fan system.

For high-temperature southern pine kilns, eight-bladed 76-inch props are available that can take from 40 to 50hp motors. This combination turns out very high performance kilns capable of drying 2-inch lumber in 12 to 13 hours with remarkably narrow moisture distributions.

There has been a great deal of controversy over the horsepower required for quality drying. The size of your motors is strongly influenced by such factors as species, desired drying time, production goals, lumber grade and end user, lumber thickness, sticker thickness, fan spacing, the presence of crossouts, the plenum width, and door height. If you ask me to design a kiln for your application, I take each of these factors into consideration. The right horsepower for your application is different from someone else's. Unfortunately, there are no hard and fast rules to determine what that horsepower should be. That is based on experience. Also, technology changes.

For example, 20 years ago 15 hp and 20 hp fans were the recommended sizes for high-temperature kilns. Today 25 hp and 30 hp fans tend to be the norm. As we have seen, 40 hp is not unheard of. Over the years we have found that higher horsepower and its additional air volume helps narrow the variation in air velocity top to bottom within the kiln, as well as across the package width. As do wider plenum chambers. This is true for all softwood kilns and I suspect hardwood as well.

This text is aimed at helping you live with and get enhanced performance from your existing system. To purchasers of new softwood kilns, I can offer the following guidelines, by no means etched in stone. All use 72-inch six-bladed props.

Conventional Kilns

Ten hp minimum on single track kilns, 15 on double. All internal cross shaft motors. I want to see a minimum of 700 fpm with an average 800 to 850 fpm. Temperatures should peak around 200°F.

Accelerated Kilns

This is an interesting kiln style with temperatures normally approaching 200 to 210°F. I have put the same fan system in these kilns as I have for conventional kilns under the assurances that the internal motors could take the heat and have been disappointed by chronic motor failures. As such, external drives with 15 hp motors on single tracks and external 20 hp motors on double tracks are recommended.

High-Temperature Kilns

These kilns operate at temperatures above the boiling point of water. Use a minimum 20 hp on single tracks, 25 hp on double tracks with eave-mounted drives, although I personally prefer 25 hp and 30 hp, respectively. For through-the-roof drives, add an extra 5 hp per fan. As air velocity increases, so can temperatures, with proportional reductions in drying time (figure 7.8). Air velocity minimums of 900 fpm, averages in the 1100 to 1200 fpm range, would be expected, but do not look for it with the minimum horsepower values noted.

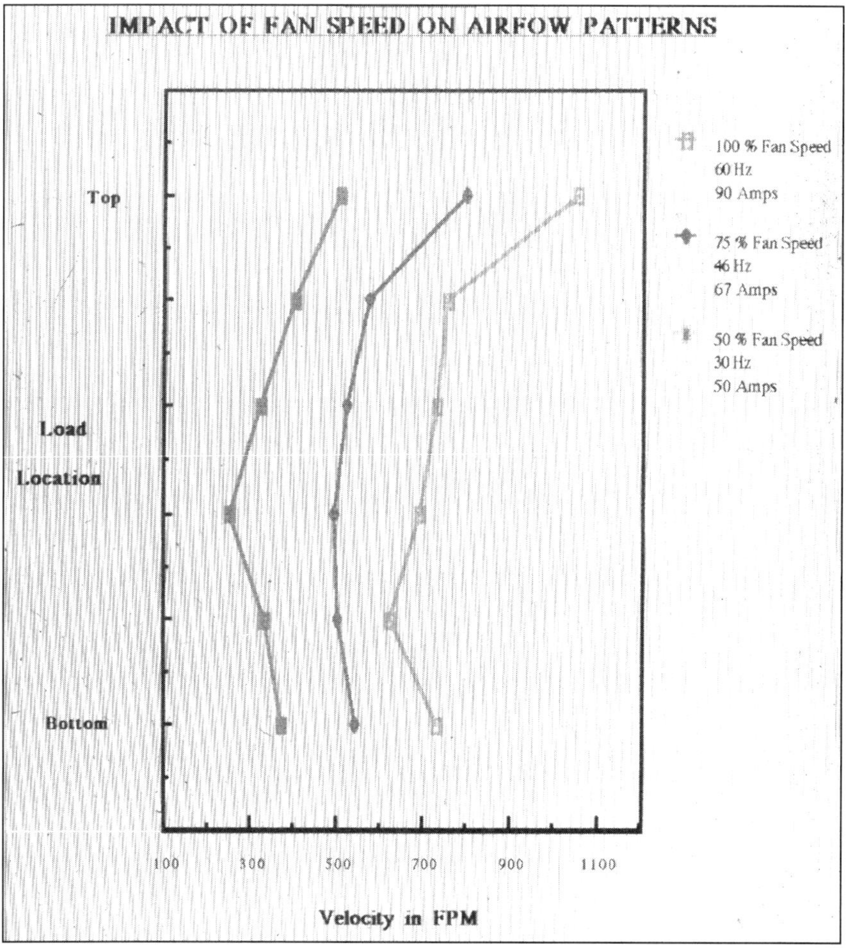

Figure 7.8 Influence of air speed on airflow uniformity top to bottom on the second track. Note how the differences widen as the speed increases.

Variable-Speed Motor Drives

We have known for decades that reducing fan speed after achieving fiber saturation point (FSP) is one way to reduce energy usage without affecting how fast we dry. The ability to reduce fan speed by varying line frequency has been available for many years. Simple two-speed motors have been available much longer.

Historically, the application of variable-speed motors required the interaction of an operator. Someone had to be on hand to turn the switch that reduced fan speed. This was a serious drawback to the application of this technology, which has been overcome today through the application of computerized controls.

This technology was originally introduced, and remains widely used, in the hardwood industry. It has been far less successful for softwoods. To understand why, it is helpful to understand how the technology works.

In drying softwoods we always aim to load lumber that is as green from the saw as possible into the kilns. Without antifungal dips, this is the best way to control stain problems and reduces the problems with variations in beginning moisture content due to air drying, which occurs as lumber waits to go into the kiln. Sapwood moisture contents of softwood lumber may vary from a little over 100% MC up to 250% for certain species. Heartwood moisture contents may range from 30% MC to 100% MC.

The vast majority of this water is held within the cell cavities as free water. The removal of **free water** is very rapid and occurs through **capillary flow**. Its removal is highly dependent on air speed. Because of this, during the early periods of our schedules, we always want to run the fans as fast as possible. The removal of free water continues down to a moisture content of around 30% MC. This 30% threshold is termed **fiber saturation point (FSP)**. Below FSP we begin to remove water from the actual cell wall or fibers of the wood. This is termed **bound water**, since the water is bound into the cell wall of the wood. Moisture removal here is by a very different mechanism known as **diffusion**, which is the migration of moisture from areas of high to areas of low concentration. Diffusion is very slow and is largely unaffected by air speed. It is below this FSP threshold that we are able to reduce fan speeds.

So why does this work well for hardwoods and not softwoods? Many hardwood species are routinely air dried prior to kiln drying simply to reduce the amount of time the lumber must occupy a kiln. In many cases, this lumber is entering the kiln at or below FSP, yet still requires 4 to 12 days to reach the desired moisture content. The period of time during which fan speeds can be reduced is relatively long.

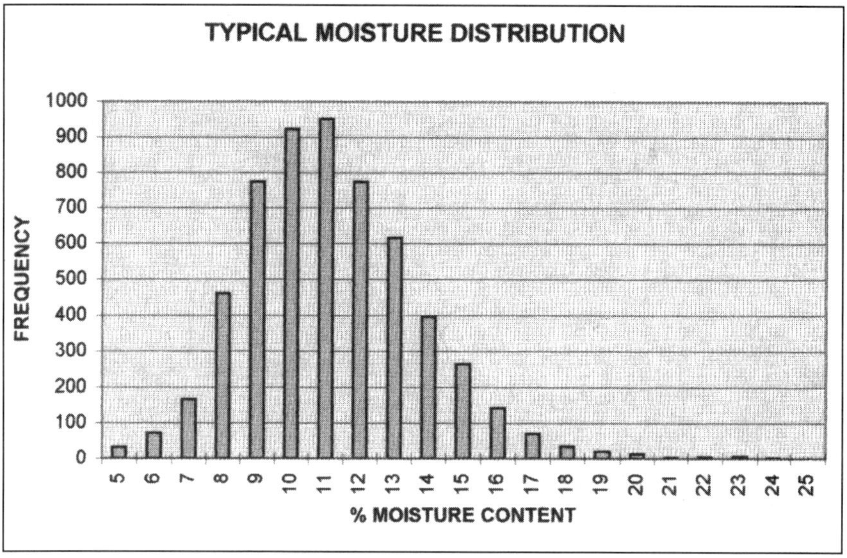

Figure 7.9 Typical moisture distribution for softwood dimension lumber having a standard deviation of 2.5. At what point in the drying cycle could you safely reduce fan speeds and be assured that all of the lumber was below fiber saturation point?

Now, think about that green softwood in the last kiln charge you ran. At the end of the charge, when you went into the kiln with your moisture meter to determine if it was dry, was it all the same moisture content? Assuredly not. Even if the majority of a load checks in the 12% to 14% range, it is not uncommon for you to hit a board or two that reads 25% or more (figure 7.9). This is the reason we have not been very successful with variable-speed drives for softwood when used to save electrical energy. There is simply too much variation in moisture content in the final product. We also have no way to reliably determine when all of the lumber is below FSP before reducing fan speeds. Although all of the lumber at or below FSP will continue to dry as though nothing is happening when the speed is reduced, those that are above FSP will slow down. This widens moisture distributions. The first time you use this drive will likely be your last, at least during the main drying schedule.

Although you cannot expect significant energy savings with variable-speed drives on softwood species, there are still several good reasons to use them.

1. We can maximize airflow by controlling motors to a constant amperage. On startup, cold, heavy kiln air is harder to move and requires more energy. As the air heats and becomes lighter, amperages fall off

by 10% or more. Controlling to a set amperage allows us to reduce the initial strain of startup and increase air speed as air becomes lighter. This allows us to increase the potential of existing motors. In some instances it allows us to use a smaller motor and still deliver plenty of air for drying.

2. The VSD allows us to use the motor as a brake to quickly slow fans for faster fan reversals. Although a minor benefit, this makes efficient use of our drying time. On restart, motor speed slowly increases to reduce or eliminate initial startup power surges. We extend the life of belts and lessen the strain on windings, bearings, and sheaves.
3. On high-grade lumber, requiring equalizing and conditioning, it is easier to maintain set points by reducing fan speeds during those portions of the schedule. Air moves in a dry kiln due to the static pressure built up by the fan system. High static pressure means high air speed, but it also draws in more unconditioned outside air into the kiln environment. Reducing the static pressure through reductions in fan speed reduces this influx.
4. By dynamically balancing our fans, we can often spin them at higher speeds without worrying about slinging a blade. The VSD will help us accomplish this and still remain within the maximum load capacity of the motor.

Although we may not save much money on electrical energy in our softwood kilns, variable-speed fan technology is an old idea whose time has come. Poorly applied in the past, modern computers have opened up a whole new world of possibilities for its use. It works because the technology finally exists to make it work.

Getting the Most Out of Your Existing Fan System

Why do we want to maximize the air speed through our lumber? Simple. For many softwood species and a few soft hardwood species, drying time is reduced as we increase air speed. The general rule of thumb is that for every percentage increase in air speed we obtain half that percentage as a reduction in drying time. Increase air speed 20% and you should reduce drying time by 10%. Increasing air speed can be as simple as tightening up what you have now. It can be as complicated as starting over. The following species will generally benefit from increases in air speed:

southern yellow pine (loblolly, slash, longleaf, shortleaf, minor species)
Douglas fir
lodgepole pine
ponderosa pine

red pine
white spruce
aspen
soft maple
yellow poplar

In order to understand our options for improving our fan system, it helps to understand the elements that go into the efficient design of one. There are four:

1. Everything mounted within the fan arc or in close proximity to the fan must be aerodynamic in shape. Start with the motor mounts. Are the leading and trailing edges leading into the fan aerodynamic in shape? Can we do anything to smooth the flow? Making these supports more aerodynamic in shape is often the only factor we can easily alter. We have already discussed how through-the-roof fan drives cost us fan performance, but unless we want to spend a lot of money, we must live with it. Placing heating coils, their feed piping, and even sprinklers for fire protection in front of the fans impacts on performance. Yet, here again, it would be expensive to make changes now. Just keep this in mind the next time you purchase a kiln (figures 7.10 and 7.11).
2. Maintain minimum tip clearances. The tip clearance is the distance from the tip of the blade to the shroud enclosure around the fan. As a blade turns, there is a tendency for the air it is pushing to sling off the end of it. The shrouding forces the air to move perpendicular to the fan's rotation. On a practical basis, due to vibration and differential

Figure 7.10 Notice how clear this fan deck is near the fans. There are no obstructions to impede airflow near the fan orifices. This design also leaves plenty of room to move around in the fan deck to do maintenance.

Figure 7.11 Notice how the heating coils have been stood upright in front of the fans in this kiln. This negatively impacts fan performance.

thermal expansion, the maximum tip clearance should be on the order of 3/8 inch to 1/2 inch. At a recent seminar someone reported getting his clearances to 1/8 inch but had been unable to document any improvement. Tolerances this close make me nervous. Any false movement, perhaps a bearing failure, and there will be trouble. Try to maintain the recommended 3/8 inch to 1/2 inch for optimum performance.

3. Maximize fan speed. For the axial fans we use in dry kilns, the faster we can turn them, the more efficient their performance. This means that as RPM increases we get more air per horsepower. For 72-inch cross shaft fans the manufacturer's recommended maximum is 900 to 925 RPM.

TABLE 7.1 TYPICAL ENTRANCE LOSSES FOR VARIOUS TYPES OF ORIFICES

Type of Orifice	Description	Loss*
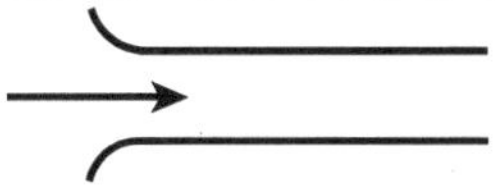	Smooth, well rounded	5
	Two-cone, 45° & 15° per side	6
	Flanged cone, 15° per side	7
	Unflanged cone, 15° per side	13
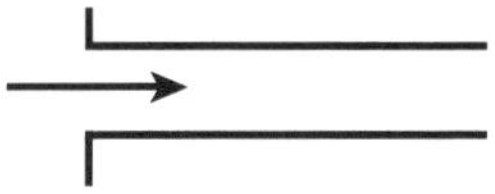	Flanged pipe	50
	Unflanged pipe	90

*Loss is given in percent of Velocity Pressure (% VP).

Line shaft fans turn much slower. If a tachometer reading indicates you have the potential to increase speeds, then discuss your intentions with the kiln manufacturer to see if he advises it. I have visited too many mills that increased fan speeds only to wreck their fan systems soon afterwards (figure 7.13).

4. Shroud design. Sooner or later as you learn more about fans, you will run into the following drawing...

On older kilns it was typical of manufacturers to roll a piece of 12-inch junior channel into a shroud, primarily for their durability. In cross section these shrouds are closest in performance characteristic to the flanged pipe shown in the figure. Note the loss values for this flange. By changing the shape of these shrouds and others, it is possible to significantly impact on the performance of the fan (figure 7.12). Ideally, we want the smooth, well-rounded orifice shown at the top of the figure.

If you are purchasing a new kiln, take a close look at the fan system of each kiln manufacturer, keeping in mind the above criteria. Question them

Figure 7.12 Fan shroud equipped with inlet cones to smooth airflow into the fan. There are rounded shrouds on the market that are even more effective in improving fan performance.

closely regarding each item and write their promises into the contract. You have to live with this equipment a long time, and a little forethought now will save a bundle of headaches later. As your kiln is under construction, visit it frequently, paying close attention to how the piping is being routed. Finally, if you must install a fire protection system, make certain it does not interfere with the fans or access to the fan deck for maintenance.

Options to Improve Fan Performance

Here is a listing of the things you might consider to improve the performance of your existing fan system.

1. Smooth all support members of the motor mount by adding a piece of angle iron to the leading and trailing edges of the supports within the fan arc. Remove or relocate all of the obstructions in front of the fans and out to the plenum chambers.
2. Tighten tip clearances if they are greater than 1/2 inch. This is best done in conjunction with the shroud change discussed in item 4 below.
3. Increase RPM up to the manufacturer's recommended maximum (figure 7.13). One major kiln manufacturer typically runs its drives around 850 RPM. With the manufacturer's blessing, increase this to 900 RPM or perhaps 925 RPM. It is possible to exceed the manufacturer's maximum,

Figure 7.13 This mill increased fan RPM without consulting the kiln manufacturer. The center fan walked off the shaft after breaking the bearing bolts, taking the fans on either side with it. Score: Kiln—three fans, one shaft, and four bearings. Mill—three days downtime plus the expense of repair and initial changes.

with its guidance, by dynamically balancing each individual prop within the kiln. A good motor rewind shop has the equipment to do this in the field, and increases to 1000 RPM are reported for 72-inch props. On sheave drives, the actual RPM increase will require a sheave change. Direct drives require a variable-speed drive. Keep in mind that if you increase the RPM of the fan, you must reduce the pitch on the blades.

4. Upgrade your shroud design to one that is smooth and well-rounded. The primary expense in this is the labor involved, so why not go all the way to the top? Fan manufacturers have told me that the radius for the inlet curve needs only be around 3 inches, although shrouds have recently come on the market with a radius approaching 12 inches. If a little is good, a whole lot ought to be better.
5. Stage start the fans. This technique only starts every other fan during the initial heat-up period. Once the heat is up and the amperage has fallen off on the motors, the fans are shut down as though for a normal fan reversal, and all fans are brought back on. This allows you to slightly overpitch your fans from what the motors can stand when they are cold and operate them at a higher average amperage when they are hot. This option is not advised for direct-drive fans due to the potential for moisture condensation on the idle fans.
6. Install a variable-speed drive. In conjunction with ammeters installed on each motor and monitored by computer control system, you can run all of the fans at maximum amperage all of the time. During the cold startup the computer would reduce speeds and gradually increase them as the loading allowed. This truly maximizes the output of any fan system.
7. Recapture air that is currently being lost. Make certain your baffles remain in place and direct air through the load after the fans are started. Weight overhead baffles that sail, firmly restrain end baffles, and do what is required to keep floor baffles from plopping over. Close the gap between concrete floor baffles and the load. If you really want to get serious, cover hinge points with belting to reduce this area of loss. Reduce or eliminate crossout openings. Make certain fan bulkhead doors and trap doors in fan decks are closed after maintenance. Close every gap you can find in fan bulkheads and fan decks. Repair holes in the kiln walls, roofs, and doors.

Be advised that the following ideas have not demonstrated any measurable improvement in kiln performance.

1. Smoothing the interior walls of the kiln by installing sheeting over the purlins does not increase airflow. Likewise, rounding the interior corners at the eaves or at the floor will not give you any benefits. Away from the fans themselves, the air speed is too low (1,000 to 1,500 fpm,

i.e., 11.4 to 17.0 mph) for aerodynamics to have much impact. They cause no harm, though, so unless they cause another problem, you need not remove them if they are already there.

2. Alternating left- and right-handed fans down the kiln length in cross shaft fan systems. I personally tested this idea back when it was in its heyday and was unable to demonstrate any benefits over having all left- or all right-handed fans in the system. There was no increase in airflow, nor was there any change in the pattern of the airflow through the lumber load, which was the original intent of this idea. The fan manufacturers sure enjoyed this fad when it was going around, though. Again, this causes no harm if already installed. There is simply no demonstrable benefit.

Maximizing Fan Performance

Obtaining peak performance from an existing fan system means maximizing and balancing fan output along the kiln length. This section deals with the steps necessary to realize the full potential of our fan system. At this point we have ideally done everything we are going to do from the above list to improve fan efficiency. Now is the time to get that last drop.

Direct-drive fan systems are not concerned with power losses from the motor to the fan. Shaft-driven systems have the potential to lose a great deal of power to improper belt tension. Belt maintenance is very important to fan efficiency, and if you hear a high-pitched squeal or squeaking sound after the fans have come to speed, you are likely losing fan efficiency from belt slippage.

Before maximizing and balancing a shaft drive fan system, check belt tension and adjust it if necessary. Replace worn belts and sheaves if they are worn. One way to reduce belt and sheave wear and smooth startups is to install airbags under the fan motors (figure 7.7). This maintains flexible belt tension and reduces the belt squealing typical of fan startups. With better control over belt tension, the variation in fan motor amperages narrows. Now that our belts are tight, we can begin.

The first step is to determine where we are. During the last charge prior to doing this work, check and record the amperage of each motor at cold startup, 3 hours into the schedule and near the end of the run. After shutdown, lock out the fans and record the maximum rated amperage stamped on the motor nameplate(s). This should give you a good idea where you are. Most operations maximize amperages with 2-inch lumber in the kiln. Keep in mind that motors work harder to push air through thicker lumber (i.e., fewer sticker openings), and you may need to maximize amperages on your thickest item to avoid motor overload. If you want to document your progress in optimizing fan performance, load a new kiln charge as for running, close the doors and vents, and check and record air velocity in the kiln before you start. Mark the spots you check with paint so you can compare point to point later.

The angle graduations stamped or cast into fan blades and hubs are not reliably accurate. I neither use nor recommend using them. We need a mechanical device for setting pitch. The most economical of these is a carpenter's inclinometer, available from most good hardware stores. The one I use has a magnetic base to which I attach the 12-inch metal rule from a tri-square. The tri-square rule functions as a wide base and, for 6-foot fans, provides a handy gauge to mark one-third radius on the blade. Another useful, though more expensive, device is a SmartLevel™. If you have Smithco fans you can use their degree gauge, although this is more useful for setting pitch than reading it (figure 7.14).

Fan pitch is measured at one third of the fan radius in from the blade tip. For a 72-inch fan this is 12 inches. At this point in the process we are simply trying to survey our existing fans to see where we are, so go through the kiln and mark a line on each blade at the one-third radius position. I use a lumber crayon for this purpose. Now, starting with fan one...

1. Face the fan assembly with the drive shaft going away from you. Position the fan hub on your right.
2. Number the blades with your crayon and bring the first blade down horizontally in front of you.
3. Set the measuring device on the top edge of the blade and level it to 0 degrees (figure 7.15). Make a mark on the fan shroud, level with the top edge of the blade. This is your horizontal mark (figure 7.16).
4. If you have not already marked the blade for one-third radius, then do so now. This is where you read the angle for setting pitch. Lay the angle device on this mark with the straightedge tight against the blade while holding the fan blade horizontal with the mark on the shroud. The straightedge must extend fully across the width of the blade (figure 7.17).
5. Read the angle and mark the value on the blade.
6. Repeat steps 4 and 5 for each blade, being careful to number each fan in sequence.
7. Repeat steps 1 through 6 for each fan in the system.
8. Record your data.

NOTE: Fan pitch is measured from the vertical axis. In order to make sense of the numbers you generate, it may be necessary for you to convert the numbers by subtracting them from 90°.

When complete your data should look something like table 7.2.

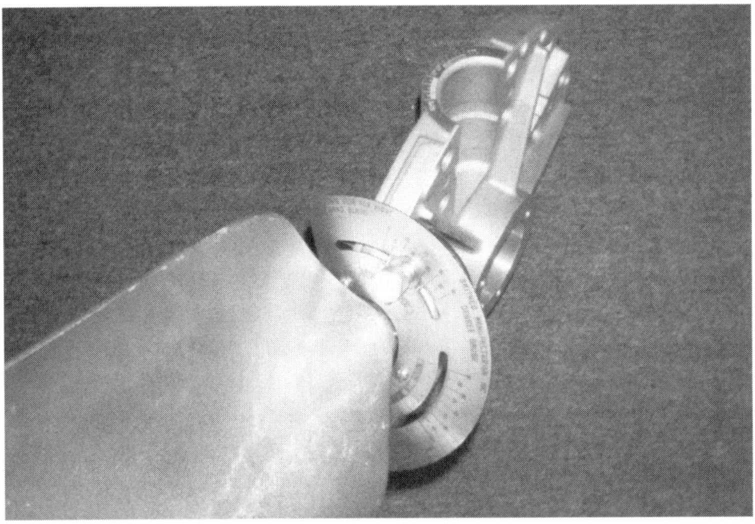

Figure 7.14 The Smithco Degree Gauge™ used for setting fan pitch on their blades. This device is fast, easy, and accurate.

TABLE 7.2 FAN PITCH AND AMPERAGE DATA FOR KILN.

	BLADE NUMBER							KILN @ 116°F	TEMP. @ 180°F
FAN #	**1**	**2**	**3**	**4**	**5**	**6**	**AVG.**	**AMPS**	**AMPS**
1	29.0	28.5	27.0	28.5	28.0	28.0	28.2	32.5	28.0
2	30.0	28.0	28.0	28.0	29.5	27.5	28.5	32.0	27.5
3	28.0	28.0	27.0	27.0	29.0	29.0	28.0	31.5	27.0
4	27.5	26.0	27.5	28.0	28.0	30.0	27.8	31.0	26.5
5	28.0	27.5	26.0	27.0	27.5	28.0	27.3	30.0	25.5
6	27.5	28.0	27.0	28.0	28.0	29.0	27.9	32.0	27.5
7	27.0	27.0	26.0	28.0	28.0	28.0	27.3	31.0	27.0
8	28.0	27.0	27.0	27.0	27.0	27.0	27.2	30.0	26.0
9	28.0	28.0	27.5	27.5	26.0	27.0	27.3	30.0	26.0
10	30.0	30.0	23.0	24.0	24.0	24.0	25.8	28.0	24.0
						ALL	**AVG**	**116 °F**	**180 °F**
					MAX	30.0	28.5	32.5	28.0
					MIN	23.0	25.8	28.0	24.0
					AVG	27.5	27.5	30.8	26.5

Figure 7.15 Checking and setting the fan pitch with an inclinometer. Bring the first blade down to 0° horizontal. Notice the manufacturer's mark for checking the fan pitch, which is one third of radius or 12 inches from the tip on this 72-inch fan.

Figure 7.16 Place an index mark on the shroud to mark the horizontal position.

Figure 7.17 The measuring device is placed on the blade at the one third of radius position and read while holding the blade horizontal.

Now, compare the pitch and motor amperage data. We are looking for two things. First, we want to minimize the variation between all blades throughout the kiln. Second, we want to maximize the loading on the motors. If we are already at or near maximum motor loading, we must be content with simply minimizing variation. When setting up a new kiln, I make certain all pitches are the same. When minimizing the variation in an existing kiln, I often find that by changing a few selected blades I can get the variation to within 1 degree and that is generally satisfactory. Ideally, though, they should all be the same. If you feel you can increase pitch, be cautious: a small change will have a large impact on motor loading. The pitch you arrive at is also influenced by stage starting and variable-speed drives. Talk to your kiln manufacturer about suggested starting pitches for your fans, then experiment

Some people like to overamp fans on startup since total amperage drops off quickly as the air warms. I have heard recommendations for up to 15% above motor rating, and it is certainly worth a try to increase output. Actual drops vary from 10 to 16% and are highly dependent on your situation. If you collected the amperage data as I described, you should have an idea how much your amperage drops off as the air warms. If you decide to do this, add a little cushion for safety. Keep in mind that you must watch your system closely after you make any changes to verify that fans do not overheat and drop out. Losing a fan has a tremendous negative impact on drying conditions. Do nothing that affects system reliability.

Typical Airflow Patterns for a Steam-Heated Kiln

The airflow patterns created within a steam kiln are influenced solely by the stacking and loading methods along with the efficiency and balance of the fan system. Unlike direct-fired kilns, airflow patterns are not generally influenced by the heat distribution system, although I have seen isolated instances where they were. We will discuss direct-fired kilns and the influence of their heating system on airflow patterns in the chapter on these kilns.

By taking firm control of our stacking and loading practices we can radically reduce the variability of airflow within our kilns. Balancing and maximizing airflow allows us to reduce this variability further. Still, there are variations inherent in the kiln design that are unaffected by these improvements. The natural variation top to bottom in the kiln, and differences between tracks in two track kilns, still remain. Figure 7.18 illustrates these variations in a typical double track kiln.

Top to bottom variation is smaller off the first track and greatest in the airflow pattern coming off the second track. Air moves in a dry kiln because the fan system produces an imbalance in the air pressure within the kiln. We measure these imbalances with the aid of a manometer, which measures the pressure differences between the two sides of the fan bulkhead in inches of water. This difference is known as **static pressure**. In a conventional kiln we might expect static pressures around 0.75 inches of water. High-temperature kilns range from 1.25 to 1.75 inches of water. The higher the static pressure, the higher the air velocity.

If we visualize the first track of a double track kiln with the air moving in a counterclockwise direction, we can imagine the first left-hand plenum chamber fully pressurized and pushing the air through the load evenly. That is why there is so little variation in the first track. Once the air clears the first track, though, it is pulled to the lowest pressure area of the kiln, which is at the uppermost corner of the second, right-hand plenum chamber. This results in a general migration of air to this upper corner, in turn resulting in the wide variation in airflow top to bottom typical of the second track. Single track kilns are renowned for their drying uniformity because the air flow in them is typical of the first track of a double track kiln.

Plenum Baffles

An interesting approach to balancing this airflow is the installation of plenum baffles in the outside walkway to split and control the airflow, a concept initiated by Gayle Bridges (figure 7.19). Plenum baffles seem to work in some kilns and not in others. I once installed these baffles at a mill that had two kilns, each radically different in heating coil design. Kiln #1

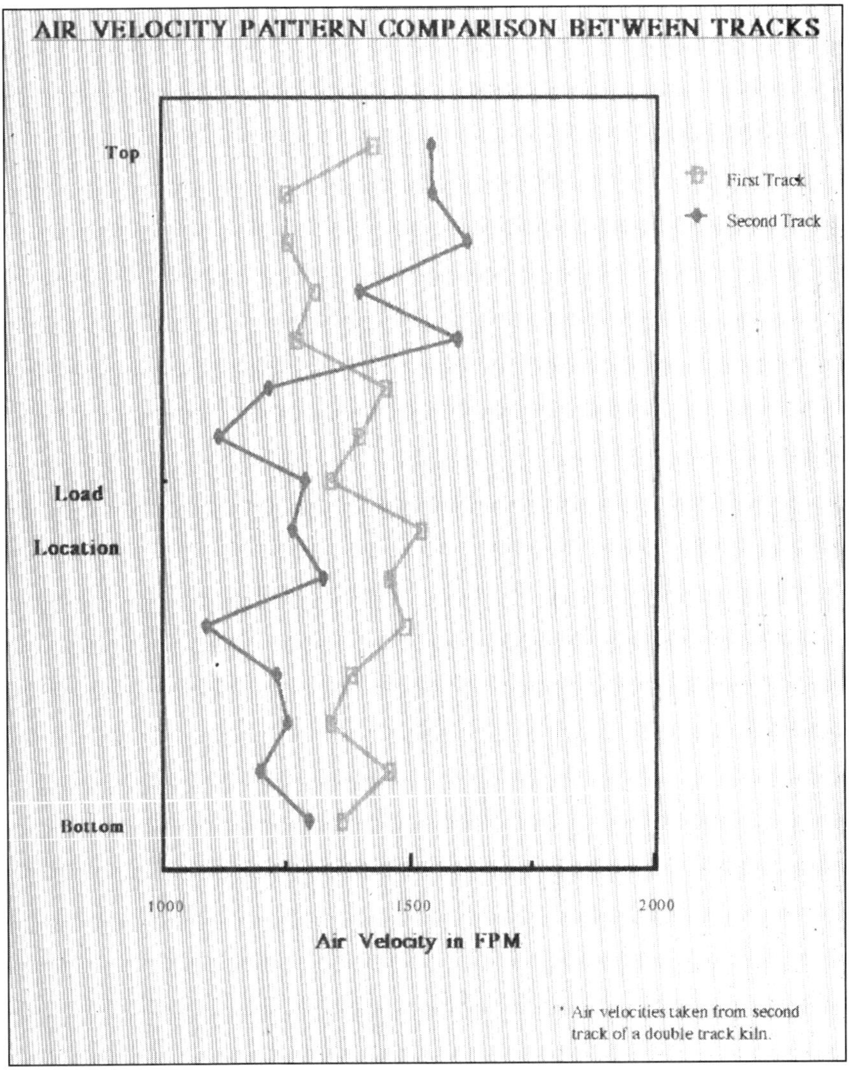

Figure 7.18 Variation in airflow, top to bottom, in a double track kiln. The first track is fairly uniform, while the second track demonstrates the shift in airspeed that occurs as air migrates to the upper layers. A single track kiln has an airflow pattern similar to the first track of this kiln.

had a center coil that could not be split top to bottom into zones. Indeed, the center coils were oriented vertically. Kiln #2 was multizoned into two zones top to bottom on the center coil. After we installed plenum baffles in the first kiln we saw a significant drop in drying time and an improvement in grade. In my enthusiasm, I recommended these baffles to other mills and found that some mills obtained good results, while others could see no change. Meanwhile, the first mill installed the baffles in Kiln #2 and saw no improvement for that kiln, although they had seen and continued to enjoy the improvements in the first kiln.

It appears from this experience that if you are able to control temperature variation top to bottom through zoning on the center coil, you will gain little benefit from installing plenum baffles. If you do not have control over the top to bottom heat variation off the center coil, you generally gain approximately one-half the potential benefit of zoning the coil. If you later zone the center coil, the plenum baffles lose their utility, although they cause no harm by keeping them installed. Due to the airflow uniformity of a single track kiln, it is doubtful that the installation of them here would have any practical utility. They are very useful in direct-fired double track kilns.

Recently, plenum baffles have been combined with multizoning, high heat release heating coils, and high-performance fans to attain some pretty remarkable achievements in drying time and quality. I am slowly beginning to warm up to them again for multizoned double track kilns.

Adjusting Plenum Baffles

If you have plenum baffles in your kiln, they were probably supplied by the kiln manufacturer. For a while a few years back they were one of the gimmicks manufacturers used to distinguish their kilns. Manufacturers' plenum baffles invariably include an upper section that is adjustable.

How do you know where to position this adjustable section? After making all the modifications you expect to make to your fan system, adjust and secure all of the plenum baffles to the same position. Whatever looks reasonable is fine. At this point we really have no idea where they should be positioned. Load the kiln with the freshest lumber you have and dry it. Pick an entire cart of lumber that is uniformly colored and put it through the planer, starting with the top.

> NOTE: You may have to test several carts in a direct-fired kiln to get a fair view of what is going on in them, due to the inherent airflow variability in those kilns.

At the planer breakdown, measure and record by layer the moisture content of as many boards as possible from every other layer as it slides past.

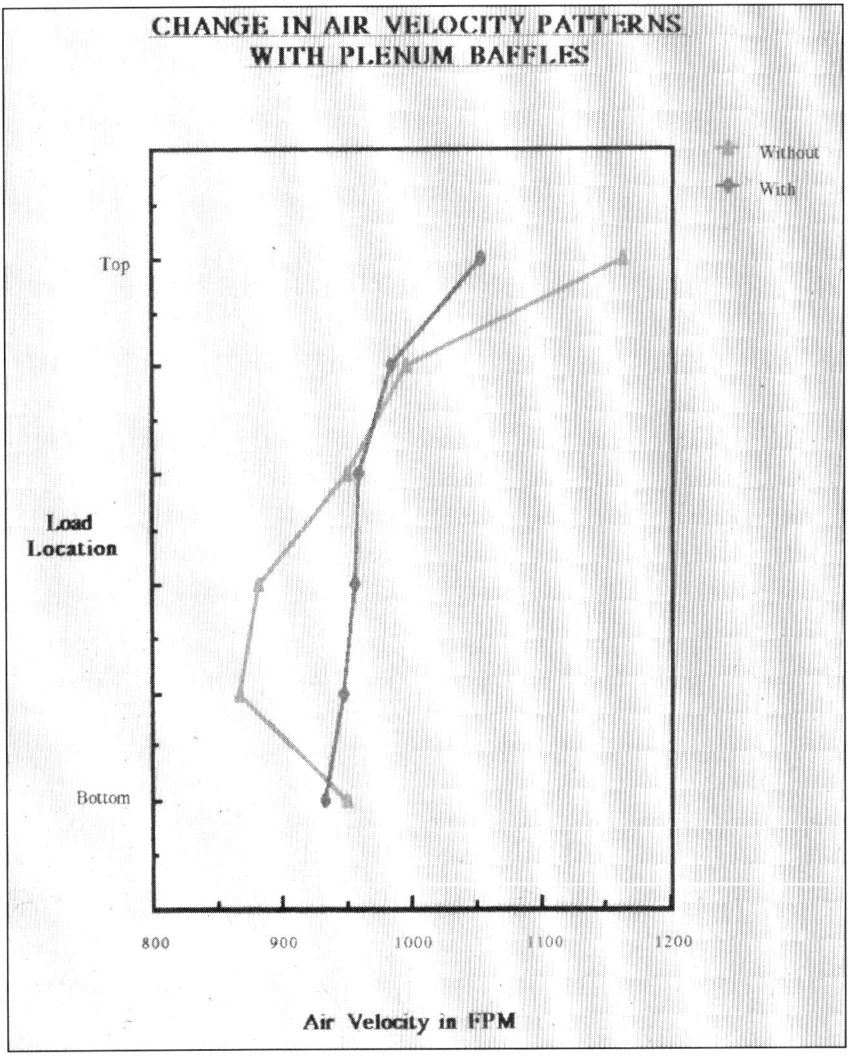

Figure 7.19 Change in airflow distribution patterns through the installation of plenum baffles. Top to bottom variation has been significantly reduced.

Average the moisture content of each layer and observe how it varies top to bottom. Based on this information, adjust the plenum baffles evenly down the kiln length, putting more or less air to the top versus the bottom as indicated by the data. Run another test charge. Again, check the moisture variation top to bottom. Keep repeating this until you are satisfied with the top to bottom moisture variation. These adjustments assume even heat transfer top to bottom from your center heating coils. If you suspect otherwise, you

Figure 7.20 A kiln equipped with plenum baffles designed with an adjustable top section to adjust flow. Once you have gone to the effort to determine where they should be positioned, weld them in place.

must correct the heat problem first, although in a few cases the plenum baffles may be used to compensate for unevenness in heat transfer.

Once you have it as close as you think is humanly possible, WELD THE HINGED SECTIONS OF THE PLENUM BAFFLES IN PLACE. After all this effort you do not want to discover six months down the road that a maintenance man thought they should be in a different position (figure 7.20).

Now, some kiln manufacturers design their plenum baffles with a second hinged section at the bottom, which is designed to swing up against the load. The idea here is to segregate the stacks vertically into zones. Have you noticed a streak of wet lumber through your stacked carts at the midpoint? This is what is causing it. Remove this bottom section.

Fan Operation and Reversing Sequence

When fans first start at the beginning of a new charge, there is a real danger of the kiln walls "imploding" or being sucked in. It is embarrassing to explain to your boss that a hole has suddenly appeared in what had been a solid aluminum wall. How does this occur? The answer involves the expansion of kiln air as it heats up after the doors are closed and its sudden contraction as it passes through the cold lumber when the fans start. All manufacturers make provisions for this in their control systems, invariably opening the vents just prior to starting the fans. This allows pressures to equalize. Some are manual and require the operator to remember to do this; *all* should be automatic. A good computerized system opens the vents before starting the fans and only starts the fans if it receives a signal that they are indeed open. I believe the exhaust side of the vents is the only one that needs to be opened during reversals, although everyone opens both. Once the fans are started, the vents should immediately close. This reduces energy loss during fan reversals.

Older kiln systems brought all fans on together in a massive power surge. Later systems sequence fans individually or in pairs to reduce surges. In addition to soft starts, variable-speed drives bring all the fans gradually up to speed to eliminate surging and have the added advantage of reducing strain on the motors' windings, belts, and sheaves. This is the wave of the future.

Most mills reverse their fans every 3 hours. I occasionally encounter 2- and 4-hour reversal times. I have even run into recommendations to reverse every hour. Other mills reverse only once—two-thirds or so through the schedule. I set my reversal times based on the charge duration. I might use 2-hour reversals for schedules running 16 to 18 hours or less, 3-hour reversals for up to 24-hour schedules, 4-hour reversals for up to 36-hour schedules, and 6-hour reversals for everything else. Reversal periods appear to have little impact on drying and tend to be a matter of personal preference. Keep in mind that you are not drying any lumber while you are reversing fans, so you want to keep their frequency and duration to a minimum. The shortness of the reversing cycle depends on how long the fans take to coast to a stop. Variable-speed fan drives shorten this time by using the drive as a brake, making a smooth transition from one direction to the other. An old friend of mine, an electrical engineer, always wanted me to "plug reverse" my fan system. In essence, this makes an immediate direction change without any coast-down allowance. Every time we met, he reminded me of his suggestion. All I could see in my mind's eye were squealing, smoking belts, twisted shafts, and blown motor windings. When he comes around, just smile and nod your head. He may have a gun.

Guidelines for Fan System Maintenance

Before doing any work on your fan system, lock out the breakers so they cannot be started. Do not forget the main breaker. Safety is foremost in any situation, and this is no exception.

The worst part about doing any fan maintenance is getting to the fans. In older kilns, it is necessary to hoist yourself through a trap door in the fan deck or squeeze between the overhead coils. Progressive kiln manufacturers recognize this problem and install gable end access doors with catwalk access in steam kilns, and internal ladders and catwalks in direct-fired kilns. This encourages frequent inspections, as does the presence of fan wall bulkhead doors. Even when I know the fans are locked out, I am always reluctant to crawl through a fan to get to the other side. Manufacturers who place bulkhead doors between every other fan should be elevated to sainthood.

Follow the kiln manufacturer's recommendations for frequency of lubrication. You should grease once a month at the minimum. Wood-fired direct-fired kilns require more frequent greasing due to the presence of abrasive ash in the air. Many line shaft systems are provided with continuous lubrication. Check these monthly.

A few mills install copper or stainless steel grease lines that extend outside the kiln. This allows the kiln operator or maintenance man to grease the kiln while it is running. Others have automatic grease systems. I do not recommend any of these. These grease lines have a tendency to work loose, depositing grease everywhere but in the bearing. Grease also hardens in these lines and makes it difficult to tell what you are actually doing. Always grease at the bearing to verify that you are doing what you think you are doing. This forces you to enter the kiln and allows you to inspect the bearing for wear and schedule replacement if necessary. Inspect fans at this time for cracks, loose nuts, and excessive movement.

Use a quality high-temperature grease recommended by the kiln manufacturer or lubricant supplier. Never mix different greases in the same bearing unless approved by the manufacturer. Different greases are often not compatible and may react chemically with disastrous results.

If fans do not have retaining washers and bolts to hold them on the shaft, install them. Fans have a nasty habit of walking off the end of the shaft and through the gable end. They might even take a few neighboring fans with them.

Install ammeters on each fan motor to monitor operation and assist balancing. As mentioned, recording fan amperages on a regular schedule helps detect slipping belts. Adapt the fan amperage log in chapter 14 to your situation. It also alerts you to the increasing friction produced by gradual bearing failure. Fan outage lights are not dependable indicators of fan outage, since they are tied into the starters, not the motors themselves.

On direct-drive fans, grasp the blade at the 12 o'clock position and push it away from you. Any movement may indicate pending bearing loss. Although manufacturers recommend that you replace motor bearings every three years, you might want to move that up a bit. The fans are balanced to what is felt is an acceptable standard at the factory, but you may consider having them dynamically balanced in the field. You will notice a marked reduction in vibration, which will help extend the life of your bearings. If you notice that your fans always stop in the same position, have them balanced.

Spin the fans. Listen and look for anything unusual.

Inspect fan blades for cracks. Examination of broken fan blades shows that some have a tendency to break one-third of the radius from the hub. It is sometimes possible to detect hairline cracks at this position upon close examination. Do this on a routine basis.

In new kilns, after changing a fan, or after resetting the pitch, check the torque on fan U-bolts, bushings, and retaining washers. Do this after having run the kiln for 2 to 3 hours at drying temperatures. Check them again after the first charge. Check them again after the first week and again after the first month. After that, check the torque twice a year. Be careful not to overtorque, particularly the bushings. Due to the differential thermal expansion between the hub and bushing, you will always be able to take up a little more. Each time you do drives the bushing farther into the hub. Eventually, you will have nowhere to go, and the fan may walk off the shaft. This is not a good thing.

Follow the manufacturer's torque recommendations. Nothing is more dismaying than taking one last nudge on the wrench and hearing a pop as the hub cracks. Never exceed the manufacturer's specifications, no matter how little you feel you are tightening the bolts. Watch out for bolts with bad threads. They can fool a torque wrench. If your fan uses stainless steel U-bolts to hold the blades in position, wipe off any aluminum shavings that might have gotten into the threads while inserting them. This will reduce spalling of the threads and nuts. I like to use a wrench that allows you to dial up the desired force in foot/pounds and makes a clicking noise when adequate force is applied. Avoid wrenches that force you to read a dial or have pointer indicators. They are inaccurate and slow you down.

Never assemble a hub and bushing on a shaft by applying WD-40™ or NeverSeize™ to the shaft. It must be dry. The fan will walk off the shaft long before you might enjoy the convenience of taking it apart.

Make certain your fans are running in the center of their shrouds. Unless the fan is at or near the middle of the shroud, you are not getting everything out of it.

Check clearance between blade tips and fan shrouds. Clearances should not exceed 1/2 inch or allow the blade tips to touch. A broken shroud occasionally slips down into the path of the fan blade, not breaking it, but slowly wearing it down. I have seen blades reduced in length by 2 inches or more without breaking. Once you repair the shroud in these situations, replace the fan. Tip clearance is critical to efficient fan performance. Keep a spare fan on hand in case you lose one. I would not bother to assemble it until you need it. Keep it in its box on a shelf and assemble it on the fan deck. Early in my career I assembled a fan on the ground because it was easier, only to discover that there was no way to get it upstairs.

Always keep spare belts on hand so that they are available when you need them. Which belts should you use? Some manufacturers use matched sets of single belts. These belts give good service and are relatively inexpensive, although they are sometimes difficult to tension. I prefer multiple belts attached to a common back. However, excessive sheave wear will split them in short order, and they can also be difficult to install. Their advantage is that you do not have to worry about tensioning several belts that vary slightly in length. Another option is power bands, which are cogged belts running on a cogged sheave, not unlike an automobile timing belt. Properly installed, these belts do not slip. This is their primary advantage. But they are also expensive and require more effort to install. Sheave alignment is more critical, and they also tend to be a bit noisier than regular belts.

If maintenance has trouble keeping your belts properly tensioned, try installing air bags under the motors. The bags are then attached to a pressure regulator, which allows you to adjust tension. The primary reason for reduced belt life is improper tensioning. A properly tensioned belt should last 12 to 18 months.

While making your fan inspection, look at other areas of the kiln. In steam kilns look for signs of leaks or unusual coil movement. Direct-fired kilns always seem to lose bolts from the distribution duct. Vent linkages slip and require readjustment. Look for damaged baffles along with signs of wall panel and structural steel deterioration.

Keeping the fan system of your kiln in good repair is paramount to maintaining drying quality.

REFERENCES

1. Taylor, F. W., and P. H. Mitchell. "Can Drying Times Be Shortened Further?" *Forest Industries* (November 1987): 24.
2. Culpepper, L. G. "Coming Trends in Kiln Drying—What's New on the Horizon?" Reno, NV: West Coast Dry Kiln Association Meeting, May 1988. Published in 1989 proceedings.
3. Culpepper, L. G. "Application of Computerized Controls to Improve the Drying Operation." Seattle, WA: IUFRO Meeting, July 1989.

Chapter Eight

Humidity Control

Good judgement comes from experience,
and experience comes from bad judgement.
Barry LePatner

The humidity control system of a dry kiln is probably the simplest and yet least understood part of the entire kiln system. It is through the humidity system that we are able to bring balance to the fan and heat system to dry our lumber. While we can easily grasp how the air moves through the lumber and heat enters the kiln, how the moisture movement is controlled often comes less easily. Yet, without humidity control, drying either slows to a stop or proceeds at too fast a pace. Both conditions have the potential to degrade lumber.

To understand how we use humidity, we must first understand the term **relative humidity**. Technically, relative humidity is defined as the ratio of the partial pressure of the vapor in a mixture to the saturation pressure of the vapor at the same temperature and pressure of the mixture. Man, I wish these guys who write this stuff would learn to speak English. For our purposes, we can think of it in the following manner. Air at a given temperature is capable of holding a maximum amount of water vapor. We would say at that point that it is saturated, incapable of holding additional moisture. The percentage that it does hold in relation to how much it could hold would be its relative humidity. If it is saturated, the relative humidity is 100%. If it is holding half of what it could hold, its relative humidity is 50%, and so on.

If the relative humidity is high, near 100%, the air does not readily accept additional moisture, and evaporation slows. You notice this on hot, sultry summer days, when your skin stays moist because your perspiration cannot evaporate as fast as it is produced. At a lower relative humidity the air readily accepts additional moisture, and you can feel the cooling action as moisture evaporates freely from your skin and clothes.

In a dry kiln we manipulate temperature, airspeed, and humidity to control lumber drying rates. At any temperature and airspeed, we can speed or slow drying by adjusting the relative humidity within the kiln—raising it to slow drying, reducing it to speed drying. There are some limits, however, to how fast or slow we can do this and still maintain lumber quality. In actual practice, we do not measure relative humidity directly but use a method of dry bulb (a simple thermometer is one) and wet bulb (a thermometer with a wet cloth over it). As water evaporates from the wet bulb, it cools. Slow evaporation cools it a little, whereas fast evaporation cools it a lot. How much it cools in relation to the dry bulb temperature is an indicator of the relative humidity. The difference between the dry bulb and wet bulb temperatures is known as the **wet bulb depression**.

There are tables, called **psychometric charts**, which we use to predict what the relative humidity is, based on the dry bulb temperature and the wet bulb depression. For example, at 180 degrees dry bulb with a wet bulb of 160 degrees (a depression of 20 degrees), the relative humidity is 62%. The air is holding 62% of the total amount of moisture it could hold at that temperature. But there is more information in these charts.

Given sufficient time, all of the wood within a dry kiln controlled at a set temperature and humidity will stabilize at a certain moisture content. This moisture content is known as the **equilibrium moisture content (EMC)**. In the above example, the EMC is 7.6%. So, if I put lumber into a kiln and maintain the dry bulb at 180 degrees and the wet bulb at 160 degrees for a long enough time, then all of the wood, even the stickers and crossouts, will stabilize at 7.6% moisture content. At that point the water in the wood will be at equilibrium with the surrounding air, neither giving up nor accepting additional moisture for any net loss or gain. This EMC value is included in the psychometric charts along with relative humidity and is actually the more important number as far as we are concerned.

We do not run our kilns by setting a single dry bulb and wet bulb depression throughout the charge, although I suppose one could. It would simply take too long for the lumber to dry. In reality, we raise temperatures and lower humidity as the drying progresses in response to how well what we are drying gives up its water. These recipes for drying, if you will, are known as **drying schedules** and vary according to species, thickness, end user, and even kiln design. Different timber species are subject to specific defects, which may become apparent depending on the schedule. Other defects occur at specific times during the schedule. All are impacted by the design of the specific kiln we are running. Determining the proper schedule for each of our items and kilns requires experience and understanding of our raw material, equipment, and customers. This is the ultimate goal of our jobs as kiln operators. At this point we need to understand the function of our humidity control equipment.

Humidity Sensors

The majority of dry kilns use wet bulbs to control humidity conditions within the kiln. This method is simple, requiring only a temperature sensor (known as a dry bulb) with a cloth sleeve or "sock" pulled over it. This sensor mounts above a water box in which the water level is controlled. The sock hangs in the water and stays wet by wicking moisture up from it. The psychometric charts are well established for this type of control.

Recently, other methods for determining humidity and EMC have been introduced. They are limited by present technology to lower-temperature drying processes. To my knowledge, they have not been successful on high-temperature kilns. EMC wafer systems allow us to determine EMC directly by measuring the electrical resistance of a known fiber sample. They work similar to a resistance moisture meter. Electronic relative humidity meters read humidity directly, but the extremes of heat and humidity found in these kilns impose severe service conditions for this type of equipment. Any decision to use alternative humidity-sensing methods must be on a trial basis until proven. For sheer simplicity and low-cost reliable operation, the wet bulb is hard to beat. I will continue our discussions using the wet bulb as our humidity sensor.

Until the early 1980s it was rare to find a dry kiln with more than one wet bulb, since the circular chart recorder/controllers were only equipped with one. Today it is common to find two wet bulbs in a computerized dry kiln, primarily as a fail-safe in case one should run dry and drive the vents open. This also gives you the option to control the wet bulb based on entering, exiting, or average wet bulb. Each control method has specific uses in particular kilns, and I appreciate the option to choose when tuning my schedules. At most, you will see perhaps 1- to 2-degree differences from side to side in a kiln, but that is often enough to affect kiln performance. In wider package kilns the difference is often greater as it often is in direct-fired kilns. Wet bulb readings do not vary significantly down the length of a kiln within the same plenum chamber.

A well-designed computer control should give you the option of entering, exiting, or average wet bulb control. Should one wet bulb run dry, the controls must automatically switch to the remaining wet bulb and sound a visual if not audible alarm. It must continue to alarm until the problem is corrected by the operator. The computer should also compare wet bulb readings to their respective average entering and exiting air dry bulb temperatures to detect and alarm should both wet bulbs run dry.

Mount your wet bulbs midway of the kiln on the side walls, 6 to 7 feet off the ground and 18 to 30 inches off the wall, well into the airflow of the plenum chambers. Mount the water box parallel to the length of the kiln with the temperature sensor 1 inch off the water surface. Any higher, and the sock tends to dry out a bit when the fans blow air down on it, giving false high readings. Socks also seem to crust up and harden more. There is no apparent advantage to mounting wet bulbs next to the center coils of double track kilns. Indeed, due to their nearness to a major heat source, false high wet bulb readings will give you fits with kiln control.

Use 3/8-inch OD copper tubing to run supply lines to the water box. You may notice that the supply lines thin out and break rather quickly and must be replaced. Cold water in these lines acts as a condenser, condensing moisture from the kiln environment, which contains weak organic acids that erode this tubing from the outside in. Consider using stainless steel for the first 10 feet of this line as it enters the kiln if you are frequently replacing it. Preheating these water lines by running them along steam lines also helps by bringing their water temperature nearer to kiln temperatures before they enter the kiln. In very northern climates I often notice that the condensed water from these lines drips onto the wall purlins and corrodes the steel. In this case, I recommend insulating the feed lines to eliminate the condensation. A larger line, perhaps 1/2-inch OD, is very nice on the drain lines. Smaller lines on the drain often quickly clog with sawdust and overflow water onto the floor. Run all drain lines with gradual bends that slope evenly towards the drain area. Sags and loops in the drain form airlocks, which prevent good drainage. Tight tubing bends tend to trap sawdust and clog (figure 8.1). Do not run this water out onto the floor to be evaporated. It eats up the concrete and corrodes the bases of support columns.

Each time you enter a kiln after a charge and pass the wet bulb troughs, reach up and feel the socks. They should be moist and pliable. If not, change them. At minimum, change the socks in conventional kilns once a week, and in high-temperature kilns every other charge. To get more wear out of your wet bulb socks, toss them into a box after use and run them through the laundry with your gym socks when the box is full.

Supply lines need some method for regulating and observing water flow. A shutoff valve to regulate flow is mandatory. Try to obtain needle valves rather than larger ball valves for finer flow control. There are several sight glass flow meters on the market that permit you to measure and regulate water flow (figure 8.2). They are handy and recommended.

Figure 8.1 Typical wet bulb installation. Avoid sharp bends in the drain lines to prevent clogging and subsequent overflow problems.

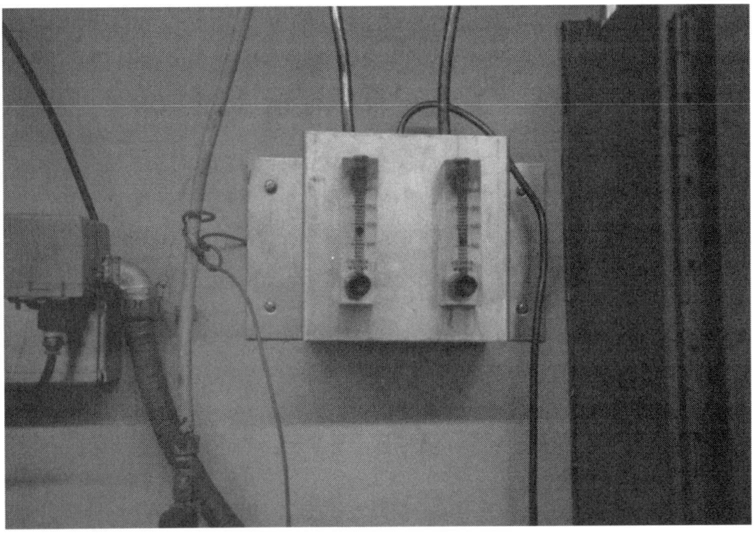

Figure 8.2 Needle valves or similar devices are helpful to control water flow to the wet bulbs. You only need enough water flow to keep the sock from drying out.

Vent Systems

Most kiln manufacturers mount vents on the roof, although a few place them on the side walls. In older line shaft kilns the vents are often all in one row, directly over the fan system. Due to the interior baffling around the fans, these vents alternate as exhaust and intake vents down the length of the kiln. There is only one actuator or air motor in these kilns (figure 8.3). In newer conventional and most high-temperature kilns the roof vents form two rows, one on each side of the fan system. One row acts as an exhaust and the other as an intake. I prefer them mounted farther down on the roof, out over the plenum chambers rather than up near the fans. The exception is in Coe direct-fired dry kilns. They typically have a power vent arrangement tied into the recirculation blower of the heating system.

In multiple-vent systems, the connecting linkages may be inside or outside of the kiln. Mounted outside, the individual vents are linked ideally with a solid or hollow rod, which connects to a pivot joint on each vent lid. I have seen vents connected by cables and even chains but find these are hard to adjust evenly along the kiln length. Use rods only, even to connect the actuator or air motor to the first vent in the row. They will prevent the vents from being blown wide open in a hard wind. Internal mounted linkages usually consist of two shafts, which extend the length of the kiln on each side under the vent rows. To open the vents, the shaft is rotated. I have a preference for this style of vent, since the vapor stream does not have to exit the kiln at a 90-degree angle to the airstream.

Figure 8.3 Vent arrangement for a line shaft dry kiln. Looks like the controller may be having problems. You should keep sufficient spare parts on hand to make repairs as soon as malfunctions are discovered.

Until a few years ago almost all vent systems were pneumatically actuated. Today electronic actuators are common and are used in conjunction with electronic heat control valves. Since I like to have total control of my drying conditions, I insist that all vent actuators, either pneumatic or electronic, be fully modulating. By that, I mean they should be able to function repeatedly over their entire operational range. If the computer says they should be at 50% open, then that is where they should be.

Many older pneumatic actuators function well only as on/off devices. They should be replaced with newer models, particularly if you are making the transition to computerized controls. Typical is one instance I recall when I spent two days attempting to adjust the control span on a set of these vents to control by computer. Even with bricks in the manufacturer-supplied counterweight box, the best we achieved was a 2-pound span. The vents began opening at 7 psi and were fully open at 9 psi. However, they did not begin to close until the pressure fell below 3 psi. Simple on/off control was the only real alternative. Computers demand more. This particular actuator has been upgraded by the manufacturer through the addition of a compression spring, which gives the air motor a little extra to work against. This is an improvement, but there are better devices on the market.

Simple on/off vent control does not give good humidity control. To maintain control around a given temperature with these devices, the vents constantly cycle open and closed. The wet bulb likewise cycles above and below the desired temperature. Rather than controlling at set point, you actually control within a range of ± 1 to 3 degrees of set point. This is not acceptable for high-quality drying, since the constant cycling breaks up your drying momentum.

The primary advantage to all pneumatic vent systems, aside from their simplicity, is that they fail closed. This reduces the likelihood of excessive degrade should they malfunction. The electronic actuators have the potential to fail open. This is their only drawback. Other than that, the electronic actuators being mounted on kilns today give excellent control with the right control system.

Adjusting Vents

The vent system must vent moisture evenly. This means that all the vents in a row must open together and the same amount. In addition, both vent rows must open the same amount when compared to each other. Variations in opening along the kiln length and from side to side contribute to moisture variations in the lumber (figure 8.4).

Figure 8.4 Typical roof-mounted actuator arrangement. All vents should open evenly down the kiln length and from side to side.

The standard maximum vent opening is one third of the vent dimension. For example, if your vents are 18 inches square, they should open 6 inches. Fully modulated vents are not bound to this and may be adjusted higher, since the controller will adjust the opening as needed. If you are short on venting area, you may want to increase the opening to its maximum opening. If your direct-fired kiln tends to warp the top layers of bundles, try adjusting the openings down a little each charge. Stop when you begin noticing extensions in drying time.

Adjustment Procedure for Roof-Mounted Vent Actuators

Tools needed: (one) 6-inch adjustable wrench, (one) 6-foot folding rule or tape rule, and (two) open-end wrenches to fit linkage bolts.

1. Force the vents open using the control system. Make certain you have full travel on the air motor. Take the air line loose at the air motor and work the vents open and closed by hand as you check for broken hinges and connections. Make repairs before proceeding.
2. Loosen the linkage between the last vent and the air motor so all vents are down. Now, go to each vent and adjust it to a fully closed position by loosening, repositioning, and retightening its connection to the linkage rod. Continue until all vents in the row are adjusted to the fully closed position.

3. Reattach the linkage to the air motor and reconnect the air line. Using the rule, measure the distance from the underside of the vent cover to the top front edge of the vent opening. Take this measurement at the widest point.
4. Adjust the linkage to the air motor until the opening is one third of the outside square dimension of the vent (a 21 × 21-inch vent will have an opening of 7 inches). For computer-controlled systems, I adjust to the maximum allowable opening.
5. Check each vent in the row to see that the openings are the same.
6. Detach the air line and check that the vents seal when closed. If everything is tight, reconnect the air line and move over to the second vent row, beginning with step 1 again.

NOTE FOR STEPS 5 AND 6: Deterioration and uneven sagging of roof panels is a common reason for uneven vent openings. Look out for this on flat-roofed kilns. Unless the sagging can be corrected, you will have to live with the variation in vent openings.

Adjustment Procedure for Internal Vent Linkages (Shaft)

Tools needed: (two) open-end or box wrenches for adjusting nuts, (one) regular screwdriver, (one) pair of Channellocks, and (one) stick or tape rule.

1. Open vents to full 100% position. Inspect the vents for broken hinges and repair any that are found. Make certain you have full rotation on the vent shaft by removing the positioner cover and checking visually. If not, adjust the span adjustment on the positioner until the shaft fully rotates.
2. Beginning at the end nearest the positioner, adjust the vent to the desired opening, based on vent dimension. Measure this distance from the underside of the vent cover to the edge of the vent opening at its widest point. Move down the line until all vent openings are the same. For computer-controlled systems, I adjust these vents to the maximum possible opening.
3. Cycle the vents closed and check for tight closure. Again, be sure that the positioner is rotated fully closed. If not, adjust the zero adjustment on the positioner until closed.
4. Repeat steps 1 through 3 for the second vent row (figures 8.5 and 8.6).

Figure 8.5 Internal view of a shaft-driven vent system. All adjustments are made internal to the kiln, which keeps personnel off the roof.

Figure 8.6 Internal shaft-driven vent system. Exhaust air does not have to turn before leaving the kiln.

Venting Methods

Up until about 10 years ago, only two venting methods were available to the kiln operator—conventional venting and exhaust-only venting. A third method, demand venting, has become available through the introduction of computer controls.

Back in the days when everyone was running line shaft kilns and using pneumatic recorder/controllers, conventional venting was the only game in town (figure 8.7). As we have seen, these kilns had a single vent row, alternating exhaust and intake vents along the kiln length, and opened with a single air motor. Later cross shaft kilns necessitated the installation of two vent rows with separate air motors. Using the same controllers, these air motors were linked together with a single control signal and operated in tandem. Conventional venting is characterized by opening both vent rows (intake and exhaust) the same amount at the same time when venting is required (figure 8.8). In conventional kilns this was the logical approach. The only way to exhaust moisture-laden air is to push it out with drier, cooler air.

When high-temperature kilns came on the scene, another vent option was developed. This was termed exhaust-only venting (figure 8.9). When venting is required, only the exhaust side vent row opens to vent moisture. High-temperature kilns are able to use this technique because they literally boil water out of the wood, converting liquid water to steam. No makeup or dilution air is necessary to vent this moisture. It is leaving the kiln, vent or

Figure 8.7 A fully modulated external pneumatic vent actuator. This unit has very fast and accurate response to signals.

no vent. Running a high-temperature kiln with conventional venting pulls in too much cold air, which drives the dry bulb temperatures down in the kiln and wastes energy. A few older high-temperature kilns continue to use this venting method. If you have one, give serious consideration to converting it to exhaust-only venting.

Once the drying schedule format for computer controls became flexible enough to accept any schedule for any species, experimentation followed. Schedules were adapted for either venting method. One of the distinct advantages of computerized controls is their ability to ramp set points up or down gradually over time. Previously set point changes were abrupt, involving fast, radical changes from one set point to the next. By combining conventional venting methods (opening both vent rows) with gradual wet bulb set point changes, dry bulb set points were easier to maintain, and drying time reduced. Quality also seemed to improve. Still, it was sometimes a little difficult to control set points with both vent rows opening the same amount.

Enter demand venting. With this method, once the computer determines that venting is required, it begins by opening the exhaust vents only to bring the wet bulb under control. With the exhaust vents open 100% and the wet bulb still too high, the computer begins opening the intake vents up to their maximum level of 100%. In actual operation, the intake vents usually modulate at some level well below this to give very tight control over

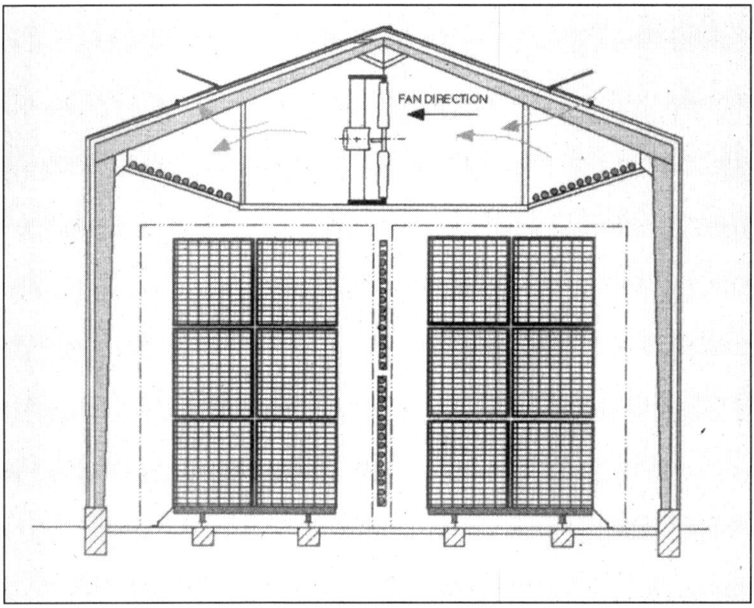

Figure 8.8 Conventional venting opens both vent rows at the same time.

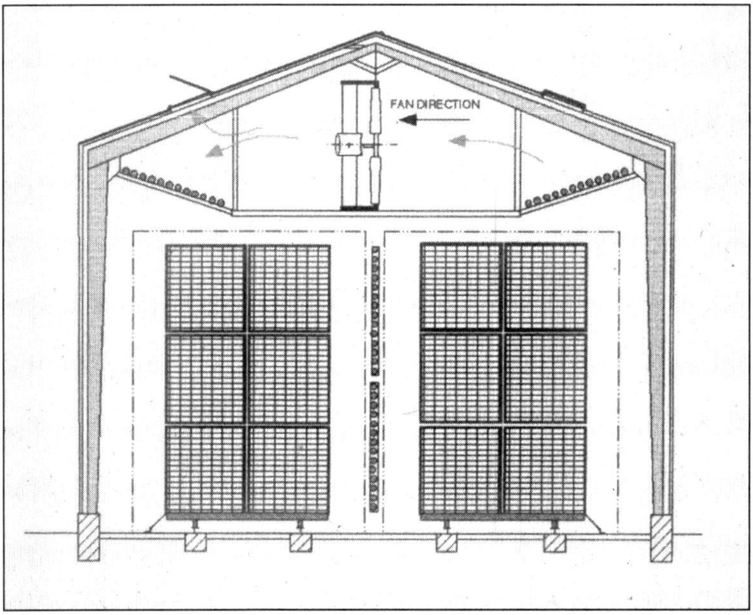

Figure 8.9 Exhaust-only venting opens only the high-pressure side of the vent system. It is used on high-temperature and direct-fired kilns.

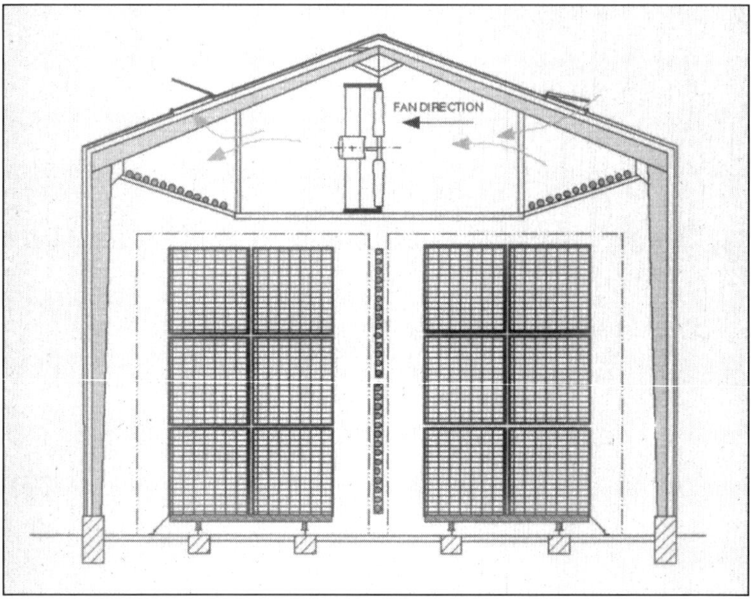

Figure 8.10 Demand venting opens the exhaust side a full 100% before modulating the intake side open to control set points.

wet bulb set points. This technique prevents overventing, vent cycling, and saves energy. I highly recommend it (figure 8.10).

Steam Sprays

While the vents are used to lower humidity levels in a dry kiln, the steam sprays serve to raise it by injecting live steam into the kiln environment. Steam sprays are mandatory for high-quality drying of boards, finish, shop, export, and specialty grades (figure 8.11). You will likely never use it for dimension grades, common boards, decking, and timbers unless you have a customer with specific needs who is willing to pay for the extra time and effort required.

The steam sprays consist of a control valve, distribution line(s) with directional nozzles, and a drain line to remove condensed water in the distribution line. As with all components of the control system, the control valve must be fully modulated across its entire output range, never on/off. The distribution line should be a minimum of 4 inches in diameter with the nozzles spaced evenly along its entire length. I usually pipe a piece of 3/8-inch OD copper from the drain end(s) of the spray distribution lines to the outside to keep this line free of water. There is no need for a valve on this line.

On low-pressure kiln systems (to 20 psig) you can use steam directly from the supply header. On high-pressure systems you may want to consider installing a desuperheater to bring the pressure down to 20 psig before injecting it into the kiln.

Figure 8.11 Steam sprays are necessary for running conditioning and equalizing schedules.

High-performance kilns normally use high-pressure steam for heating. Their pressures range from 100 psig to 200 psig, with the largest percentage in the 120 psig to 150 psig range. Even at the lower pressures, we want to get the pressure down further, though 95% of the heat in steam is released as it condenses on the lumber surface. Spraying high-pressure steam into the kiln in enough quantities to affect humidity often raises the kiln dry bulb temperature, and we lose control of kiln conditions. Yet somehow we must introduce steam to condition.

One solution is to desuperheat the steam by reducing its pressure and mixing in water to bring its temperature down. The first desuperheater I ever saw was simply a water line stuck into the steam spray system. When the sprays came on, water sprayed into the steam flow. There was little control over the quality of the low-pressure steam produced, and this system did not work well. We quickly abandoned it.

A good desuperheater uses a pressure-reducing valve (PRV) to maintain pressure in a separate low-pressure steam header. Most operate with a header pressure of 20 psig. As steam enters the header through the PRV, water, if it is needed to bring down the steam temperature, is sprayed into the flow stream. The header is tapped by the spray control valve to supply steam spray to the kiln.

A second option is to build a water level controlled trough in the kiln in which steam coils are submerged. A control valve on the coil feed lines regulates steam flow into the coil, boiling the water. This method supplies totally saturated steam to the kiln and has the added advantage of eliminating the excessive water treatment costs associated with spraying live steam, since all condensate is returned to the boiler.

Direct-fired kilns require a steam generator to supply conditioning steam. Another option for direct-fired kilns and high-pressure steam kilns is the use of a DeMystifier™ sold by SII. This system uses filtered water under very high pressure, which is introduced to the kilns as a fine mist. This mist flashes to steam at kiln operating temperatures and reportedly works well, although at several mills water quality has been an issue in keeping them in operation.

Vent Heat Exchangers

Over the past 10 years or so, vent heat exchangers have been introduced into the market. These units are typically designed with four large ducts (18 inches to 30 inches in diameter), arranged in pairs, and mounted on the fan deck. These ducts are then ducted to a roof-mounted heat exchanger. Warm, moist air is drawn out of the larger pair of ducts and through the heat exchanger, preheating the ambient air, which is injected into the kiln via the smaller ducts. Some pretty remarkable energy savings have been claimed for this equipment, which is not inexpensive. In general, the farther north your operation is, the greater your benefit. Mills in the Deep South have reportedly not seen much benefit. The mills in this region that I am aware of have all discontinued their use. More northern mills often swear by them (figure 8.12).

With this equipment, you may see an extension of drying time from what you are presently experiencing, particularly in the summer. This is due to the reduction in vent capacity over what you may already be running. Furthermore, I have always been uncomfortable with such large ducts being mounted on the fan deck and suspect that there is significant disruption of the airflow characteristics of the kiln. At any rate, installing these units in your kiln will change them, be it for better or worse. My recommendation, if you are considering this equipment, is that you obtain a *complete* installation list and phone the dozen mills nearest to you. You want to know what tangible benefits the mill experienced when the equipment was installed. It would not hurt to discover how much influence the person you are talking to had in the decision to make the purchase. Someone more remote from the decision-making process may have an altogether different opinion.

Figure 8.12 Vent exchanger, which preheats cool intake air with warm, moist exhaust, mounted on the roof. These systems are useful in northern climates to relieve strain on the boiler but have not been uniformly successful in the South.

Humidity Control and Scheduling Strategies

When setting up a drying schedule, my goal is for the kiln to run in absolute control for as long a period as possible. We wish to be in control from the moment we reach our first set points, anywhere from 3 to perhaps 8 hours into the schedule. The reason for this is simple. Drying defects occur at specific segments of the schedule. If a defect presents itself, I generally know where in the schedule to look to make changes to correct it. However, if that segment of the schedule is out of control, I have no way of knowing what to do to correct the problem. The best I can do is guess. When I go into a mill to correct a drying problem, my first job is to install schedules that run in control, within the capability of the equipment. I really do not care very much if the first schedule is wrong, provided it runs in control. I will know how to fix any problems that arise on the next charge.

Precise humidity control is fundamental to quality drying. I want to be able to modulate vents and sprays across their entire operational range without swings in set point. Now, this does not mean that I expect to establish just any set points I please and have the control system and kiln follow them. I must be reasonable to establish my expectations within the ability of the system. At the same time, tight control capabilities allow me to probe the limits of the system capabilities. I do not want to push the system beyond its ability, but I do expect it to perform at or near its limit for optimum performance. Having said that, let's examine some of the additional uses you might put your humidity system to.

Operating the sprays during warm-up reduces your time to set point by perhaps 15 minutes. However, if you are spraying during warm-up to close up surface checks, you will find that the checks will be wider and deeper when you reach charge end than they were when you started. Worse, since the kiln and all of its components are relatively cool, massive amounts of water condense on them, hastening the corrosion of all steel components such as heating coils, feed and drain lines, traps, structural steel, and rails. Do not spray during this period.

Toward the end of the charge, the sprays will help maintain humidity conditions. By maintaining set points, they smooth the transition into equalizing and conditioning and help narrow moisture distributions. However, this prevents the use of wet bulbs to indicate dryness, as is commonly done in high-temperature kilns, since you are controlling the wet bulb level above where it would normally fall as the lumber dries.

Equalizing

We need to start this discussion by defining a few terms. **Equalizing** is a drying step that is added at the end of a schedule whose sole purpose is to reduce the variation in final moisture content throughout the entire kiln. This is different from the equalization that naturally occurs within a board as the high moisture content center of the board equalizes with its surface. **Conditioning** is a drying step, usually run after equalization, whose sole purpose is to eliminate the internal stress that builds up in lumber as it dries. I find a great deal of confusion with these terms within the softwood industry. Many salespersons will say they want conditioned lumber when they really mean equalized. There is a huge difference.

At the conclusion of our drying schedule there will be a variation in moisture content between adjacent boards, within units, within carts, between tracks, and within the kiln as a whole. We describe this variation with the term **standard deviation**. Standard deviation is a dimensionless term, and it is meaningless for describing moisture variation unless we also say what the average moisture content is. Fortunately, we do not have to do the statistical number crunching necessary to generate the number. It is often done for us with the newer electronic moisture meters on the market. These meters are either handheld or in-line at the planer.

The standard deviation tells us that 95% of our lumber will fall within a moisture range that is ±2 standard deviations of the average moisture content.

Let's say that we have a charge of 2 × 8s that was run through an in-line moisture meter at the planing mill as it was dressed. In this example, the lumber run had an average moisture content of 11.4% with a standard deviation of 2.5. Ninety-five percent of our lumber will be found between ±5% MC of 11.4% MC, or within the range of 6.4% MC to 16.4% MC. Since these 2 × 8s are being stamped KD-19, this is a little low for this product, and we can gently nudge the moisture content up on subsequent charges. Looks to me like we could bring the average MC on this lumber up 2.5% MC and still remain within the rules.

Obviously, we want to make the standard deviation value as low as possible. For dimension lumber, 2.5 to 3.0 is about as good as you can or *want* to attain. By this I mean that for dimension lumber we want to get the standard deviation as low as possible without resorting to adding an equalization step. Equalization steps take time and cost money to run. You do not want to add this expense on top of your production costs unless the customer requires a tighter standard *and* is willing to pay for it.

Finish, shop, specialty, and other high-grade products require narrower moisture distributions, and standard deviations of 0.75 to 1.0 are the normal targets. Almost all of these products have an upper moisture limit of 12% MC, so if you target an average of 10% MC with a standard deviation of 1.0, that will put 95% of your lumber in the 8% to 12% MC range. You are getting a lot more money for these products, so you can afford the extra time it takes. That quality is what the customer is paying for.

The equalization step is added at the end of a regular drying schedule (table 8.1). During this step we set equilibrium moisture content (EMC) conditions that are 2% below our target moisture content. If we are targeting an average moisture content of 10% MC, then we would set EMC conditions for 8% MC. This sets a floor below which moisture contents will not fall. Everything that is below 8% MC at the outset of equalization will gain moisture to 8%. Everything above 8% MC will continue to dry down. The step is run long enough for the moisture distribution to be within an acceptable range. This requires experience based on your species, product, and equipment. I usually start out at 2 to 4 hours, run a charge, check it, and then adjust the time accordingly.

TABLE 8.1 SET POINTS FOR EQUALIZING SCHEDULES

Dry Bulb Set Points	Target Moisture Content (% MC) 9	10	11	12	13	14	15
	Wet Bulb Set points						
150	125	130	133	136	139	140	142
160	136	140	144	147	149	151	152
170	146	151	154	157	159	161	163
180	157	162	165	168	170	172	173
190	168	172	175	178	180	182	184
200	179	183	186	188	191	192	194

Try to equalize at the highest dry bulb temperature at which you can control the wet bulb. This allows you to equalize in the shortest period of time. You must determine what is best for your kiln through trial and error. One point is important here: Your kiln must be capable of holding set points within very narrow tolerances. Looking at the chart, it is easy to see that for any dry bulb set point, small variations in wet bulb shift you into the next target moisture content. This is one instance where the variable-speed drive can be useful. By reducing fan speeds to 25% or 30% during equalization and conditioning, it is easier to attain and maintain set points. In the absence of a VSD you might try shutting off every other fan on the system, although this may cause you problems with internal cross shaft fans as moisture condenses on the idle motors. Keep in mind that it is very difficult to attain the higher EMC values on this chart.

When do you begin equalizing? The rule for equalizing says to begin when the driest boards reach a moisture content of target minus 2%. This rule was developed for hardwood kilns running sample blocks to monitor drying progress. How do we know when the driest board is at 8%? We don't, really. No one cuts moisture sample boards for softwood kilns. In-kiln moisture meters give you moisture contents of selected boards, none of which are the driest boards in the kiln. Softwood mills running equalizing schedules usually begin them at a preset time, usually a set time before the lumber is expected to be dry. Computerized systems with some method for determining average moisture content and those kilns with internal moisture meters begin equalizing when they reach a certain average moisture content. What that moisture content is can only be determined by trial and error.

When I was a lumber inspector I supervised a mill that practiced its own form of equalizing. The kiln operator simply shut off the heat and closed the vents one hour before the lumber was dry, keeping the fans running. He allowed the kiln to run in this manner for two more hours. This seemed to work well for them and they rarely had moisture problems. The lumber looked good, too.

Conditioning

Conditioning schedules are run at the very end of the charge, after the equalizing step, if there is one, and serve to relieve drying stresses or **case-hardening** within the lumber. This is important for lumber that will be remanufactured. Unrelieved stresses cause lumber to warp out of shape and pinch saws as the lumber is ripped or resawn. This is a particular concern for furniture and millwork companies. In general, conditioning is not practiced on lumber whose average moisture content is above 12%.

You do not want to condition routinely to relieve stresses. Like equalization, this step takes time. Unless the customer has specific needs for conditioned lumber *and* is willing to pay for it, you should not bother. Be aware that your lumber salesman may start demanding conditioned lumber from you. Often this is part of an advertising or sales promotion that is oriented towards establishing an aura of perceived value in his customers. If he is selling dimension lumber, he probably means he wants it equalized. Here is your chance to educate him. You must determine what his objective really is in order to deliver.

The rule for conditioning requires that the EMC conditions be set at target plus 3%. For example, if your target moisture content is 10%, set your kiln EMC conditions to 13%. Table 8.2 allows you to select the drying conditions for your target moisture content directly. The 3% allowance has been figured in for you.

TABLE 8.2 SET POINTS FOR CONDITIONING SCHEDULES

Dry Bulb Set Points	Target Moisture Content (% MC) 6	7	8	9	10	11	12
	Wet Bulb Set points						
150	133	136	138	140	141	143	144
160	144	146	148	150	152	153	154
170	154	156	158	160	162	163	164
180	164	170	169	170	172	173	175
190	175	177	179	181	182	184	185
200	185	187	190	191	193	194	195

Again, notice how small variations in wet bulb temperature at any dry bulb set point affect the target MC. Tight control of kiln conditions is vitally important for successful conditioning. As with equalizing, it is best to condition at the highest temperature your kiln is capable of maintaining. Complicating this is a tendency for the dry bulb temperatures to rise as you spray to raise humidity. This is a particular problem at the higher target moisture contents and dry bulb temperatures. Again, a VSD is handy for attaining conditioning set points as is disabling every other fan as described with equalizing.

When do you begin conditioning? Start at the end of equalizing or following a cooldown period after your regular charge. Lumber does not condition above 212°F so you must get your temperatures down before starting if you are running a high-temperature schedule. Run the fans with the vents open or open the doors to cool the kiln and lumber down. You might even want to push the lumber out for a few hours. Keep in mind that you cannot condition lumber above an average target moisture content of about 12% MC.

Condition until stresses are relieved. Determine this by cutting samples from the lumber and doing the traditional tine test. In this test a short, 1- to 1 1/2-inch-long section is removed from a board, about 6 inches from the end. Using a band saw, make parallel cuts 1/4 inch to 3/8 inch below both board surfaces, stopping about 3/4 inch from the end. Remove the center core. If the outsides pinch together, the lumber surface is case-hardened, and you need to continue conditioning. If they bow apart, you have gone too far and **reverse case-hardened** the lumber. There is no remedy for this. If the tines remain straight, conditioning is complete. Congratulations. After a while you will know how long you need to condition and may skip using these tests.

For both equalization and conditioning steps, it is useful to ramp between set points if your control system will permit it. Taking 30 minutes or more to ramp from your ending schedule set points up to the equalization set points and from there to conditioning set points uses the higher moisture content boards in the kiln to help you raise humidity. Slowing these transition periods also minimizes the tendency for the dry bulb temperatures to overshoot set points as steam sprays start dumping in steam.

The Australians take a different approach to equalizing and conditioning. The softwood lumber they saw is very prone to warp due to unrelieved drying stresses. Rather than condition within the kiln, they build a separate conditioning chamber for each kiln. Once lumber completes its drying cycle in the kiln, it is pushed into the conditioning chamber to finish. This releases the kiln for drying and allows the mill to condition as much as is necessary without impacting on production. Not a bad approach if you can afford it.

Today, the softwood markets are increasingly competitive. Mills now look toward the specialty markets, where moisture is critical. Recent years have seen them develop the special schedules they need to compete in this field. There are no hard and fast solutions that apply to every situation, only guidelines. When you get right down to it, the only right solution is the one that works for you. It is up to you to discover it.

Humidity System Maintenance

Check vent linkages at least once a month. With the air signal on, take the air lines loose at the motor and work the vents back and forth. If you have a helper and two radios, simply have him stroke the vents at your command, checking their position at the 0%, 25%, 50%, 75%, and 100% positions. Pay close attention to linkage connections at the vents and at any intermediate supports. If the linkages must be replaced, use tubular aluminum. This is light, easily installed, and durable. Avoid using cables, chains, or other flexible linkages. They make it difficult to adjust vent openings because there is too much play in them. A few manufacturers use chain to connect the air motor to the first vent. Unless there is a chain attached to the far end of the vent row, the wind will catch these vents and blow them open. Being directly in the hot exhaust, these end restraining chains rust and fail quickly. Replace the connecting chains with rods.

Check the air motor where it attaches to the roof. Air motors with on/off controls tend to loosen their bolts and pull out, lifting the motor rather than the vents.

Air motor diaphragm failures are sudden and unpredictable. Keep a new diaphragm on hand at all times and monitor your charts daily for evidence that a failure has occurred. Charts from kilns with one vent row out tend to override wet bulb set points in one fan direction. They are almost in control in the

Figure 8.13 Occasionally take a look at the exhaust coming from your kiln. Uneven flow may be an indication of another problem.

other. Air line leaks also appear this way. To isolate a leaking air line, first take the air line loose from the controller. Place your finger over the air line opening on the controller. If the pressure gauge for the air line does not go to full pressure, the problem is in the controller. If it does go to full pressure, connect the air line, and take it loose at the air motor. Place your finger over the end of the air line. If the pressure gauge does not go to full pressure, the air line leaks. If the gauge shows full pressure, you have either a bad diaphragm or a broken connection at the air motor (compression nuts often split).

Use nylon tubing, rather than copper, for air lines. It is available in 250-foot and longer rolls, which permit long runs without intermediate couplings. Copper tubing normally comes in 50-foot rolls requiring couplings for runs over 50 feet. Couplings tend to work loose and leak over time, particularly when allowed to vibrate against the kiln. At 25% the cost for copper, nylon is inexpensive to replace as needed. It is also much easier to work with, requiring only a sharp knife to cut to length. It uses conventional ferrule connectors. Nylon is the ideal tubing material for multizone dry kilns, which have high requirement for tubing runs. Nylon's service life varies from two to five years, depending on installation. Run it in a separate tray, away from the kiln walls and steam pipes. Some varieties become brittle during extended sunlight exposure. Use covered trays in high-exposure situations and purchase tubing with the highest available UV resistance. This tubing is available in sets of bundled tubes that hold up longer to outdoor service conditions.

The drain on spray headers tends to become clogged with scale over a short period of time. If the drain is equipped with a shutoff valve, or even a trap, remove it. After you have cleared any blockage, install a 3/8-inch copper tube in the drain opening and run it outside the kiln and down. Although a small amount of steam is lost when spraying, you can see that it is working. I have been badly burned by water sloshing from these distribution lines between kiln charges. If you notice a lot of rust accumulating in the kiln at the end towards the drain, it is likely that the drain is clogged. With the distribution header full of water you will be spraying it and not steam into the kiln environment.

REFERENCES

1. EMC data derived from tables published in the *Kiln Operator's Manual for Eastern Canada,* by Cech, M. Y., and F. Pfaff. Ottawa, Ontario: Eastern Forest Products Laboratory, 1977.
2. Equalizing and conditioning guidelines taken from the *Dry Kiln Operator's Manual,* by Rasmussen, E. F., Agricultural Handbook No. 188, Forest Service, GPO, 1961.
3. Hart, C. Arthur. "Kiln Overheat When Conditioning Lumber." *Forest Products Journal* Vol. 40, No. 4:9–14.

Chapter Nine

Steam-Heated Kilns

We always admire the other fellow more after we have tried to do his job.

William Feather

Steam-heated kilns offer the greatest potential for heat control of all drying systems. They have a reputation for producing quality, bright lumber with high-grade recovery and narrow moisture distributions. For high-grade domestic and export markets, where all these factors are important, it is the drying method of choice.

Steam kilns require boilers to generate steam. This is a significant capital investment. Even so, it does not pay to take shortcuts while selecting the boiler for your operation. You will have to live with the wisdom of your decision for a long while and face it every day. The initial investment in a bargain boiler is quickly overshadowed by mounting operating and maintenance costs. Your boiler must be engineered to perform reliably for many years, but above all it must be of sufficient size to handle your needs. There is nothing more frustrating than trying to run a drying system with inadequate steam to do the job. You never seem to catch up. Worse, inadequate steam means inadequate heat to achieve and maintain control set points. This directly affects quality. Mills in this situation are forced to overfire their boilers to maintain production. This shortens the life of any steam-generating system. On the other hand, steam is steam. As a kiln operator, I do not care where or how I get my steam, provided it has sufficient pressure and there is enough of it. If I am also responsible for generating my own steam, then that is another matter. I need a boiler that doesn't give me problems. I have enough to worry about just drying lumber.

For mills drying with high-temperature kilns, running a boiler means running a wood-fired boiler. The high energy demands of these kilns place fossil-fuel boilers out of reach of all but a few. Wood waste is cheap, readily available, and often difficult to dispose of. Using it as fuel solves disposal problems and brings the mill tangible benefits in the form of dried lumber. More mills have also begun investigating the advantages of using a portion of their boiler capacity to generate power. As purchased power costs rise, this option becomes increasingly attractive, although not economically feasible for all.

Thermal Oil–Heated Kilns

Some kilns heat using thermal oil as a heat-transfer medium. In operation, these kilns control and dry similar to steam-heated kilns, although they use pumps instead of control valves and do not require traps (figure 9.1). The primary advantage of these systems is the high temperatures they can attain. Steam at 150 psig has a temperature of almost 366°F. Thermal oil generally operates in the 450°F range with only the head pressure of the pumping system. For this reason, they are the logical choice for kilns designed to operate using ultra-high temperatures and air velocities (300°F dry bulb/2,000 fpm air velocity).

Thermal oil has seen wide use in Canada due to the extreme risk of freezing in conventional steam-heated systems. Canada also has relaxed requirements for boiler engineers on these systems. Since they operate at reduced pressures, operators are not required around the clock. Even a low-pressure boiler in Canada is required to have a certified boiler engineer on site 24 hours a day. If you plan to run conditioning and equalization schedules, you will have to install a separate steam generator for that purpose or use another method for humidification.

Another drawback with thermal oil kilns is that even heat transfer from the heating coils is not as easy to achieve as it is with steam-heated kilns. Imagine a steam coil that has been pressurized with 150 psig steam. If you measure the surface temperature of that coil at any point across its face, the temperature will be very nearly the same, just under 366°F. It is very easy to get even heat transfer with steam. With thermal oil you are flowing a fluid that gives up heat as it travels through the heating coil, and the surface temperature of that coil depends on where you are measuring it. I have seen coil designs for these kilns that specify inlet oil temperatures of 500°F and outlet temperatures of 400°F. In order to minimize the variation in heat transfer across the face of a given coil, you must parallel the piping so that the hotter oil piping is offset by the cooler returning oil piping. It can be done, albeit imperfectly. Steam always provides more even heat transfer than any other heating medium.

Figure 9.1 Thermal oil–heated kilns require pumps to circulate the oil in the heating coils. This system is multizoned.

Thermal oil systems are uncommon in the United States. Their operating costs offer no advantage over boilers, and there is the problem of cleanup and disposal in the event of an oil spill. Unless you are on a power trip to dry faster, the additional costs of this technology give you no tangible benefit. The technology exists to accomplish all of our drying goals with steam and perform a superior job.

Steam Kilns

Let us begin our discussion of steam systems at the boiler and work our way back to the kilns. Although boiler operation is beyond the scope of this book, it is useful to have an understanding of the interaction between boilers and the drying system.

Boilers

All boilers must be operated as slaves to the kiln system; that is their primary purpose for being. If we must make adjustments to how we run our kilns other than ramping valves open on startup and using an energy management system to assist the boiler when it is in trouble, then something is wrong (figure 9.2). I want all the steam a boiler can produce up to its rated capacity. If I need more than it is designed to generate, then we need to build another boiler.

Figure 9.2 The boiler exists because we require steam to dry our lumber. It should function as a slave to the system, not dictate to it.

Be careful with the energy management systems on dry kilns. They can mask problems at the boiler by making it appear that the boiler is maintaining steam pressure while the actual amount of steam being delivered is less than what should be available. This is occasionally intentional.

As we heat up a boiler from a cold start, the energy going into it heats the water to the boiling point. At atmospheric pressure, for every **BTU** (British Thermal Unit) we transfer to each pound of water in the boiler, its temperature rises 1°F. This heat is known as **sensible heat**. We add heat. The temperature goes up. That makes sense. This energy is not particularly important to us from a drying standpoint (figure 9.3).

Once the water reaches boiling point (212°F at atmospheric pressure) something different begins to occur. As we add heat, the temperature does not rise. Adding 970 BTUs to each pound of water turns it into steam, also at 212°F. This is known as **latent heat** because it is hidden to us. This latent heat is the energy we use to transfer heat from the boiler to the kilns to dry our lumber. As steam condenses back to water within the dry kiln heating coils, it releases this latent energy to the kiln environment. The hot condensed water, known as **condensate**, is returned to the boiler to repeat the cycle.

Pressure (psig) is not as important to us as is the flow (PPH) that is available. Pressure determines the size of the steam main, the control valves, the traps, and the amount of heating coil we put into the kiln. This is engineered at the outset for our kilns, and we simply want the pressure

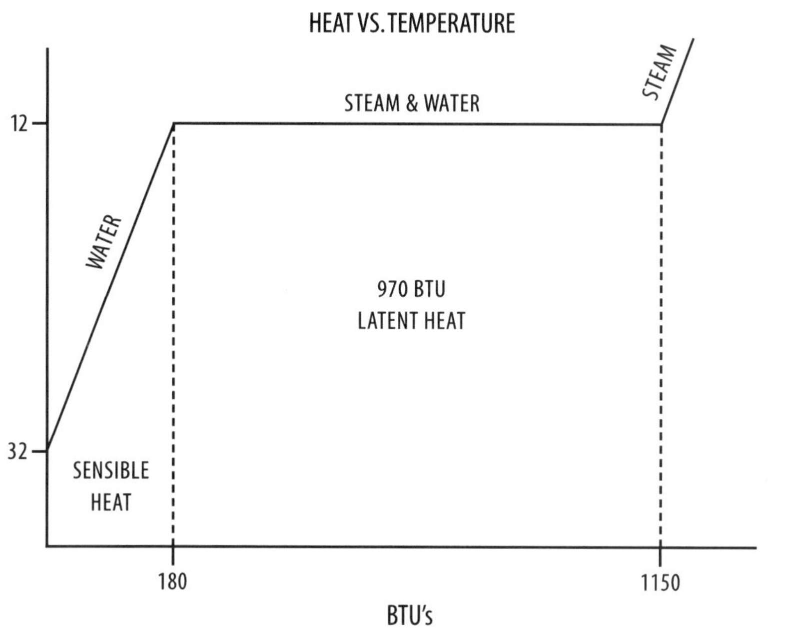

Figure 9.3 The relationship between sensible heat and latent heat.

to remain within operating parameters for our particular system. Pressure simply affects how quickly we can transfer heat within a given heating system. Flow determines the total amount of lumber we can dry, since it represents the total amount of heat available to us. Heat is what dries our lumber.

A few notes on energy management systems. All function on pressure, although a few also monitor flow mainly to keep the boiler within any permit limitations. Always mount the pressure sensor for these systems at the boiler and upstream of any pressure-reducing stations. Depending on the length of the steam main, there will always be a pressure drop from the boiler to the kilns. This pressure drop is unimportant to us at the kilns, and the harder we draw steam the greater this drop will be. If we measure the pressure at the kilns, we will often find the kilns cutting back needlessly when we need the flow the most. Set the first pressure at which the kiln begins to cut back at a low enough level to make the boiler work. If it is not at high fire while you are limiting the valve opening under your steam management system, then the boiler is not giving you everything it is capable of.

Steam Mains and Headers

Coming off the top of the boiler drum, we first encounter a main stop valve to control the flow of steam into the steam main that delivers steam to the kilns. If there are two or more boilers on the system, then each boiler will have two stop valves to allow personnel to work on one boiler while the other(s) remain(s) in service.

The steam main travels towards the kiln and must always slope away from the boiler. The main is not secured to its supports but rest on or hangs from them, being designed for free movement from thermal expansion. Occasionally they will be secured at one point to control the direction of the expansion that occurs as the pipe heats. Along the route you should find various turns and even loops, which serve as expansion joints to limit total longitudinal movement. Every 100 feet or so a drip leg is installed to collect condensate that forms along the pipe length. This condensate is trapped directly back into the condensate return line, which parallels the main and slopes toward the boiler.

If the steam main slopes towards the boiler, or if a drip leg trap should fail, the pipe will fill with condensate. Steam moves very fast in a steam main, on the magnitude of 90 to 100 mph. As this steam flows over this condensate, it forms waves not unlike the way waves form on a body of water as the wind blows over it. If one of these waves should completely fill the pipe diameter, then that wave, as a slug of water, is propelled to the end of the line at steam speeds. This is termed **water hammer**. The force of this slug of water as it reaches the end of the line is sufficient to break pipe joints, destroy valves, and wrap pipe supports into pretzels. Proper design of the steam main, and the maintenance of its various components, is important to the long-term running of your dry kiln operation. There is a sound that is sometimes heard while the kiln is standing idle between charges. People often refer to this as water hammer. It sounds as if a troop of gremlins is attacking the pipes with little hammers. In actuality, what you are hearing is the sudden collapse of steam bubbles within the condensate stream. Although the noise is annoying, it is harmless.

When opening a main stop valve, or any valve for that matter, the valve must be opened slowly to allow the piping down the line to heat up and expand slowly. Opening these valves quickly will have a result similar to what is encountered with water hammer. If I have only one boiler and am bringing it up after a cold startup, I do so with the main stop valve open. I also open a drain valve at the farthest end of the steam main line, often on the main steam header to the kilns. This allows the steam main to heat up slowly with the boiler and helps purge air that is detrimental to the operation of our kilns. Once the valve has been opened or closed to its full position, always back off a turn to keep the valve from freezing open/closed. You will appreciate this the next time you want to use it.

The main steam header serving the kilns must always be equipped with a drip leg, trap, and a blow-down valve. Mount the trap at ground level for ease of maintenance.

Control Valves

Most steam control valves for dry kilns are pneumatic. Occasionally I find steam valves with electric positioners. There is one important difference you should be aware of: Pneumatic valves always fail closed, whereas electric valves have the potential to fail open, which can cause you to override set points. Knowing this, both valve types are otherwise acceptable to dry kiln operation. In cold or humid climates, the electric valves are often preferred.

Control valves must give tight shutoff against full steam pressure to prevent temperatures from overriding set point. On pneumatic valves, this means using a valve actuator that operates through a 6–30 psig signal range. Valves with 3–15 psig signal range do not give full shutoff at line pressures above around 100 psig unless they are equipped with a positioner. The valves must control steam flow throughout their span for accurate temperature control. Never use on/off valves for dry kiln applications, since you cannot avoid over- and underriding set points.

Most kiln manufacturers place small shutoff valves ahead of each control valve on the system. The idea is that you can work on any single control valve while the kiln is running. This is a good idea, particularly on longer drying softwood species. Safety and comfort is the deciding factor here.

A few manufacturers install bypass lines around their control valves. These lines are a convenient way to preheat a kiln before startup. On cold startups, they reduce the small steam leaks that often result from fast, uneven heat expansion in the coils. Coils can be heated slowly. In conventional kilns, bypass lines are used to maintain some steam flow to the kiln while a faulty valve is changed. If the control valve is undersized, opening bypass lines during startup increases steam flow. Indeed, this is often their primary function in kilns whose valves are undersized. In actual practice, bypass lines are a chronic source of steam leaks and offer small benefit for their trouble. If used to increase steam flow during high-demand periods, they are often forgotten and left on. This causes temperatures to override and affects grade. Do not install bypass lines. They are a waste of money (figure 9.4).

Steam leaks often occur immediately on cold startup. Reduce them by opening valves slowly a few percent per minute, or leave the valves cracked open 10 to 20% between charges. This last suggestion speeds heat-up and keeps the coils purged of condensate.

Figure 9.4 Control valve installed with a manual bypass. This adds needless expense and maintenance problems to the system.

Weld all possible steam line fittings to reduce leaks. Use socket weld fittings on pipe 2 inches and less in diameter; butt weld on all else. Screwed valves require unions on both sides for valve removal and repair. Standardize valve fittings to make them interchangeable. This allows you to keep a complete valve with fittings on hand and assembled for quick replacement in an emergency. All valves are stamped or cast with an arrow indicating flow direction. Make certain this is correct on new installations and when valves are replaced.

Figure 9.5 Control valves mounted for a multizoned dry kiln. Although multizoning was originally developed for high-temperature dry kilns, it has proven its worth for all lumber drying applications.

Feed lines from the control valves to the heating coils should ideally be run in such a way that it is not necessary to crawl over, through, or around them to do service work in the fan deck. Insist on this while a new kiln is under construction. You will appreciate it later.

I prefer nylon tubing for air lines to the control valves. It is less expensive than copper and permits very long runs without couplings, which are prone to leak. Run it in covered tray for full support and protection. On multizoned kilns with several valves, always label each valve as to zone so you can locate it easily should you suspect a problem while the kiln is running (figure 9.5).

Steam Coils

In steam kilns, controlling the flow of pressurized steam through the control valves into the heating coils transfers heat into the kiln. These coils are fabricated from pipe, around which a fin is wrapped to increase its surface area for greater heat transfer. This pipe is called a **finpipe** (figure 9.6). As heat transfers from the steam through the finpipe walls, the steam condenses and drains to the bottom of the coils as liquid condensate or water. The latent heat is released into the kiln environment. For each pound of steam condensed in the coil, a pound of condensate is produced. This condensate is removed from the coils by traps; otherwise, condensate would

Figure 9.6 Horizontally oriented finpipe installed in a multizoned kiln. Each coil is individually trapped.

pool in the coil, and no further heat would transfer to the kiln environment. The difference in heat transfer between a steam-filled coil and one that is waterlogged is on the magnitude of 1000:1.

The amount of steam finpipe you need in your kiln depends on many interrelated factors. Steam pressure, lumber species and their expected drying times, lumber thickness, kiln capacity, and the type and size of finpipe being used are important considerations. Kiln manufacturers look at all of these and more when designing the heating system for your kiln. Still, coil design is not an exact science. Experience plays a larger role than might readily be admitted. When in doubt, it is always best to take the conservative route. Additional finpipe in the initial installation is relatively inexpensive. As long as you do not restrict air flows, you can never have too much. It is much more difficult and expensive to add more later if you find yourself short. No one ever complained about having too much finpipe.

All modern kilns use finpipe rather than bare pipe for their heat systems. Older kilns may still be using bare pipe coils, and you should consider replacing it to improve kiln performance. Many finpipe styles are available using different combinations of pipe size, fin thickness, fin height, and fin material. From an economy and long-term service standpoint, schedule 40 steel pipe with 14 or 16 gauge rolled steel fin is hard to beat. It is important that the fins tightly contact the pipe for maximum heat transfer. Loose fins are little better than bare pipe. Keep this in mind when you are replacing finpipe. Never cut the fins loose when cutting finpipe to length. Tack weld the fin to the pipe on either side of the cut before cutting.

A few manufacturers offer steel pipe with aluminum fins. Stainless steel finpipe is available as well. Both are useful in specific applications. My biggest complaint about the more exotic finpipe is that manufacturers often overrate its heat transfer capability and use less footage to compensate for its expense. More often than not, these kilns are heat deficient. The exotic fin materials also offer less resistance to mechanical damage and tend to bend easily when struck. This is a concern on center coils exposed to falling lumber. Tightly spaced fin material also tends to trap and hold sawdust particles, which can become a fire hazard. I prefer standard steel finpipe for my softwood and most hardwood kilns. If you are drying oak, then stainless steel and aluminum offer advantages for corrosion resistance.

Coil and Control Zone Design

One of the fundamental things we are looking for in our heating coil design is the uniform transfer of heat top to bottom and along the length of the kiln. If there are gaps around and between coils where air is able to slip by without being heated, then we are not obtaining even heat transfer (figure 9.7).

Double track kilns are designed with two distinct sets of coils. The main heating coils are located between the two pairs of tracks. Most people refer to these as the center coils or booster coils. I consider them to be the primary heating coils for a double track kiln. A second set of coils, called overhead coils, mounts at or above the overhead fan deck over the side plenum chambers. Single track and package-style kilns have all of their coils overhead.

Figure 9.7 Coils that do not cover the entire load area result in undetected cold spots and wet lumber.

The center coils stand upright between the tracks and should fill an area roughly equivalent to the area of the kiln load as viewed from the side. At the bottom, they must begin at the top of the cart and extend to the top of the load for the kiln's full length. The individual coils must run horizontally, with a slight slope toward their drain ends to give uniform heat transfer top to bottom. Coils that sag in the middle allow condensate to pool, which reduces the heat transfer capability of the coil. It also accelerates corrosion. Baffle any space between the bottom of the center coil and the floor to prevent air from circumventing under the coil and not being heated (figure 9.9). This is often a source of wet lumber.

Occasionally one finds center coils where the finpipe stands upright. These coils do not transfer heat evenly top to bottom. At the top of the coil there is good contact between the steam and the interior of the pipe. Here we get maximum heat transfer from its surface. As steam condenses, the condensate sheets down the pipe, keeping the steam from touching the pipe interior. Heat transfer is significantly less at the bottom (figure 9.8). It is interesting to note that should you measure the surface temperature of the coil at the top and at the bottom of these coils, they will be the same. This is because 150 psig steam, for example, has a temperature of 366°F. When it turns to condensate its temperature remains 366°F. What is missing is the latent heat, nearly 1000 BTUs worth per pound, that was given up in the phase change. These coils transfer less heat at the bottom, which contributes to wet lumber in the bottom units of lumber carts dried in these kilns.

Figure 9.8 Vertically oriented finpipe with multiple coil banks sharing a single coil. Both conditions contribute to uneven heat delivery.

Figure 9.9 If you notice a wet spot in the lumber from the bottom of your carts, adding baffles beneath the center coil may help eliminate it.

The overhead coils either mount flat or at a slight angle over the outside plenum areas and adjacent to the overhead fan deck. They are mounted with a slight slope for drainage. The traditional method for controlling heat to the overhead coils places both overhead coil banks on the same control valve. As air circulates in either fan direction, it must first pass through one of these heated coils. If the kiln is venting, then preheated air is immediately exhausted, a wasteful loss of energy. Another method places both overhead coils on separate valves. This variation is known as the vent loss minimizer, or VLM. With the fans blowing in either direction, air passes through the first overhead coil without preheating, and venting occurs without undue energy loss. As the air continues through the second coil bank, it heats as needed to maintain drying conditions. This may be very useful in northern climates to limit vented heat loss. In either system, if there is a gap wider than 6 inches between the overhead coils and the outside wall, it must be closed with a baffle. Otherwise, unheated air slipping past the coil will impact drying in the bottom of your loads.

A few manufacturers mount overhead coils vertically in front of the fans, placing coil banks on one or both sides of the fans. This design restricts access to the fans for maintenance. Other manufacturers place coils directly at the fan to increase its heat transfer. Usually there is only one coil bank, which allows them to use less coil and cut costs. In practice, having the overhead heat source between the vents results in excessive vented heat loss because all vented air is completely preheated when it is exhausted. It is normal for these kilns to struggle to maintain dry bulb set points once venting begins. They benefit the most from vent-mounted heat conservation devices, which reclaim exhausted heat. The biggest drawback to this coil configuration is the negative impact of having the obstructing coils in front of the fan system. This is not a good design.

Multizoning

Until 1982, all high-temperature kilns were built with one or at most two heat control zones. Single-zoned kilns sense temperatures in one location for each fan direction and modulate steam through one valve to control set points. This places both overhead and center coils on the same valve. Since the overhead and center coils are rarely balanced to deliver uniform heat, this zoning technique commonly results in very wide variations in moisture distribution.

Double-zoned kilns are found in several variations. The first divides the kiln into two zones down its length, in effect dividing the kiln into two short single-zoned kilns. Although this helps control lengthwise heat variation to a degree, these kilns are little better than single-zoned kilns in terms of drying quality. The preferred double-zoned kiln configuration places the center and overhead coils on separate controls. This method at least permits some control over the heat to each track.

In 1980, I conducted moisture and airflow distribution studies in a two-zoned kiln divided into overhead and center control zones. This investiga-

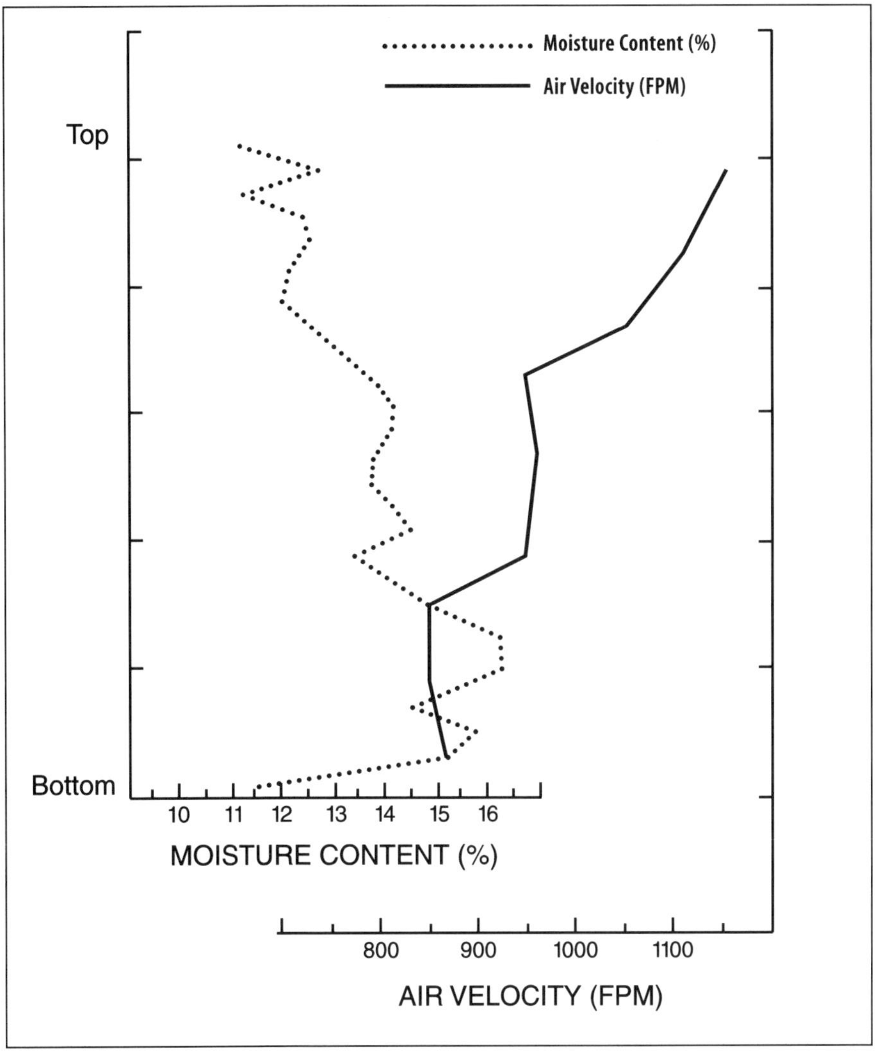

Figure 9.10 The inverse relationship between air velocity and moisture content helps us understand why tops of carts can be overdry and their bottoms wet.

tion revealed that as one went top to bottom within the kiln, moisture content increased. Air velocity data showed that airspeed decreased top to bottom and demonstrated an inverse relationship between airspeed and final moisture content. Figure 9.10 illustrates this airflow and moisture relationship within the test kiln. Several attempts to correct top to bottom moisture variation by baffling the center coils to slow airflow through the top were not successful. Being unable to control airflow effectively, we turned to heat as the control variable.

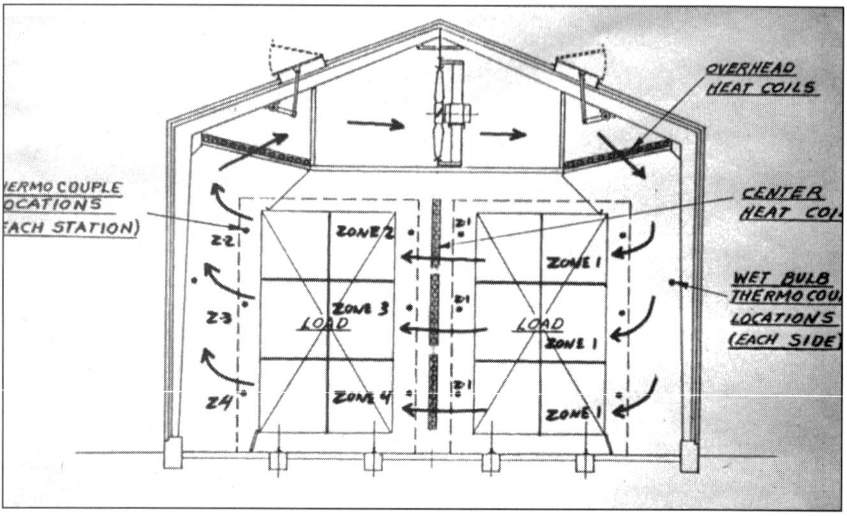

Figure 9.11 Cross section drawing of a multizoned double track kiln. The center coils are split three high with separate controls for the overhead coils.

The test kiln and the proposed kilns were loaded with three units per truck and included two crossout spaces. From the moisture distribution and airflow data, we concluded that three heating zones top to bottom with a fourth overhead were required.

Additional data from my studies of an in-kiln moisture monitor had shown wide variations in moisture content down the test kiln's length. Due to the coil configuration of the proposed kiln, we divided the kiln into three longitudinal zones (stations). This gave us two short stations nearest the doors, with a larger middle station. The result was a total of 12 control zones per kiln: (3 center coils + 1 overhead) × 3 stations = 12 zones. Figure 9.11 illustrates the coil design for these kilns.

The primary benefit of the multizone configuration is the ability to direct heat when and where needed within the kiln to maintain drying conditions. This must be combined with exiting air heat control to obtain the most benefit. The wisdom of this approach was immediately obvious on the initial system startup. Due to enhanced airflow through the top units, they achieve set point far ahead of the lower units. They cut back while lower coils continued to receive heat according to their needs. Conventionally zoned kilns allow the top coils to continue to gain heat as lower coils struggle to set point. This contributes to the overdry condition of top units in these kilns. Multizoned kilns have the potential to eliminate this problem. We will discuss the tangible benefits of multizoning in the chapter on kiln controls.

Figure 9.12 How to properly install an inverted bucket trap within a dry kiln. First, the trap is easily accessible from the ground and mounted at a good working height. There are shutoff valves before and after the trap, so we can safely break the unions for trap maintenance. The downstream shutoff valve can be closed and the smaller test valve on the trap top opened to visually test its operation. Finally, there is a dropout and blow-down ahead of the trap to remove trash. A strainer equipped with a blow-down would work here just as well. I neither recommend nor install check valves downstream of any trap.

Single track kilns, because the coils are all overhead, are almost always single- or double-zoned. Recently, a few single track kilns have been built with multiple zones down their length. The overhead coils can also be split to control heat distribution top to bottom in the kiln. This is a distinct advantage, since you are able to control temperatures along the length and height to reduce variability. The primary reason we do not see more single track dry kilns is that their relative cost per board foot of drying capacity is about 30% higher. More are being built, though, because of the flexibility and quality drying they offer.

Steam Traps

There are a startling variety of traps on the market that are available for dry kiln use. Of all of them, the inverted bucket trap is the only trap suitable to the application. It is the standard trap for dry kilns for a variety of reasons (figure 9.12). Here is what *Heating/Piping/Air Conditioning*[5] had to say in an article about traps: "The inverted bucket trap provides a combination of advantages that are not available in any of the other trap models. The operation of the trap is intermittent and provides for excellent [condensate removal] action. The inverted bucket also offers excellent resistance to wear, corrosion, dirt, and hydraulic shock. It continues to operate at full capacity on very light loads and against high back pressure. Finally, the inverted bucket trap has established its long-term ability to conserve energy for a much longer period of time than any other steam trap available today."

The most compelling reason for using inverted bucket traps is not mentioned in this article. The inverted bucket trap is the only trap capable of removing noncondensables (carbon dioxide and oxygen) from the steam system. Noncondensables reduce the heating efficiency of the steam, in effect, diluting it. Carbon dioxide forms carbonic acid (CO_2) and oxygen causes oxygen pitting on the interior surfaces of coils, feed, and drain lines, eating holes in your equipment from the inside out. Other traps (float and thermostatic, thermostatic, TLV) claim to remove noncondensables but only do so during warm-up. Once steam reaches these traps, they no longer remove noncondensables. The inverted bucket is the only one that does so. I recommend no other trap for dry kiln applications.

Never install orifice traps in a dry kiln. These traps are simply holes in a plate mounted in the drain line. The hole's diameter is determined by the average steam load of the system and is designed to pass condensate, or steam, continuously. Sizing is a problem with no satisfactory solution. Do you simply size the hole for average load conditions? The average steam demand is a far cry from the extremes in condensate load found between startup and shutdown in kilns. Orifice traps remove condensate too slowly at the beginning of the charge and waste steam at the end. They are a bad idea all the way around.

Many manufacturers install thermostatic air relief valves on their coils. Although I realize the importance of removing this air from the system, I have always had problems with these air vents. Being nonessential to the operation of the kiln (from the kiln operators' point of view), they are seldom serviced. When they fail, they blow live steam into the kiln, upsetting kiln conditions. Upon first opening a kiln, while making that first walk-through to pull back the baffles, I am listening for steam leaks. These vents seem to whistle and blow more than they should, and I always end up plugging them to eliminate false alarms. I have never observed a difference in kiln performance after plugging them. Still, the decision is up to you. To avoid any controversy over this issue, leave a little steam on the lines to keep the coils hot between charges. This will prevent the coils from drawing air into the system.

I like to install my traps in the kiln, keeping them at a convenient working level (as opposed to traps in a pit), and reducing their chances for freezing in the winter. With kiln charges so short, repairs wait until charge end. Others prefer them outside in a pit or lined up along the exterior wall (figure 9.13). I have stopped arguing about which one is "best." This is a matter of personal preference. Keep in mind that poorly maintained trap pits are difficult if not impossible to work in. You might as well have them in the kiln.

Check traps between charges by simply turning on the heat and testing each trap in turn. Do this once a week as part of your preventive maintenance program. If you have a computerized, multizone dry kiln, spot check traps every day by comparing entering and exiting temperatures through the coils. No heat rise through a coil means you have a problem, not necessarily related to the trap.

Install traps as far below the coil as possible for drainage. Again, weld all possible joints and standardize your trap assemblies. Make sure the flow arrow on the trap points toward the drain. Keep spare traps already assembled with the required nipples and unions for quick change-out. Make certain each trap has a strainer or mud leg with a blow-down valve to remove trash ahead of it. Never place check valves downstream of a trap. They have a high failure rate and serve little practical purpose. They also cause problems when they stick closed. I remove them where I find them.

Most kiln manufacturers install shutoff valves before and after their traps to quickly isolate and replace traps. There are unions on each side to speed removal. The valves are also used for checking trap operation when another valve is mounted on the test port of the trap. This is discussed in detail later. An alternative method places no valves at the traps but places one in the main condensate line to isolate the system. To change traps, the main steam and condensate line valves are closed. Traps are checked by listening to them.

Figure 9.13 Traps mounted in a pit. Occasionally they are lined up along the outside kiln wall and positioned under the control valve for the zone it drains. This allows the operator to check the two crucial components of his heating system.

The primary advantage of this system is the reduction of potential sites for steam leaks, although I prefer to be able to isolate individual traps.

Each coil must be supplied with its own trap. Tying two or more coils together on one trap contributes to cold spots within the kiln. The only exception is double-banked coils set side by side. One trap will serve each double bank with no problem.

Checking Trap Operation

Check your traps for proper operation at least once a week. If you have a computerized multizone kiln control system with temperature sensors at each coil bank, use it to check them every day. However, you must still double check these manually once a week.

To check traps through the computer system, examine the display screens or printouts that give you information on temperature and valve openings. It is best to do this during startup when the valves are 100% open and the kiln has been running at least an hour. Look closely at the temperatures on either side of each center and overhead coil. If the air temperature does not rise as it passes through a coil, then suspect a problem.

First, verify that the computer is telling the control valve to be open, and that it is indeed open. If not, determine the reason. If the control valve is open, then the trap is suspect. If it is early in the charge and the trap is accessible, shut the kiln down and test it manually. Otherwise, complete the charge and mark the lumber in the vicinity of the coil served by the bad trap for special handling. There is a strong likelihood that it is wet.

There are two recommended methods for checking traps depending on how your traps are set up. They are the visual method and the audible method.

Checking Traps Visually

This method is used on traps installed with isolation valves on the inlet and outlet sides, with a smaller test valve on the top of the trap.

1. With the heat on, close the valve on the outlet side of the trap.
2. Open the test valve on the top of the trap. Be certain its outlet points away from you or any bystanders.
3. If the trap is working, condensate blows outward, starting and stopping at regular intervals.
4. If steam blows out continuously, check the following:
   a) The trap may have lost its prime. Close the inlet valve and let the trap sit a few moments to let condensate accumulate. Open the inlet and test again.
   b) The valve may not be seating properly due to worn parts or scale lodged in the orifice. Pull the trap down and replace the kit.
5. If condensate blows continuously, then the coils are probably waterlogged. The trap will be cold. Check the following:
   a) The steam pressure may be too high because the wrong pressure was originally specified.
   b) The pressure was raised without installing a smaller orifice.
   c) The pressure relief valve has failed.
   d) The boiler pressure gauge reads low.

e) Normal wear has enlarged the trap orifice.
f) A high vacuum pressure in the return line has increased the pressure differential beyond the trap's capability.

NOTE: Occasionally I find a cold trap that begins working when struck. Go ahead and replace the kit.

6. If no condensate or steam reaches the trap, consider the following:
   a) The strainer ahead of the trap or the pipe line or elbows may be plugged.
   b) A valve in line to the trap may be broken.
   c) The trap mechanism may be worn or defective.
   d) The trap body or vent may be filled with dirt.
7. If the trap is hot but no condensate is discharged:
   a) The trap may be installed above a leaky bypass valve.
   b) There may be a leak in the blow-down valve in the condensate line feeding the trap.

Testing Traps Audibly

This method is useful on any trap installation.

1. Turn the heat on to the kiln. Using a mechanic's stethoscope available from an auto parts store, touch the metal probe end to the top of the trap near its outlet (figure 9.14).
2. Listen to the sound:
   a) A properly functioning trap makes a loud "breathing noise" as the trap fills and discharges. Often the clattering sound of the inverted bucket is heard as it drops after discharge.
   b) A continuous rushing sound has two causes. Splash a little water on the top of the trap. If it boils violently, then the trap is blowing through. The trap must be repaired. If the water sits without boiling, then the trap is still discharging cold condensate from the system. Allow time for the lines to clear, and test again.
   c) No sound indicates that the trap is dead. Check cutoff valves to make certain condensate may pass freely. Repair the trap if no restriction or blockage is found.

Guidelines for Steam System Maintenance

Finding and repairing steam leaks is often an ongoing project at the dry kiln. Not only do steam leaks introduce uncontrolled moisture into the kiln, they cost you money for fuel and water treatment. Mills with chronic, uncontrolled steam leaks often find that the boiler runs easier after replac-

Figure 9.14 Testing a trap by listening to its operation. A properly functioning trap sounds something like heavy breathing when it discharges.

ing deteriorating coils and repairing leaks. Table 9.1 illustrates the money even small steam leaks represent. Numerous small leaks easily add up.

TABLE 9.1. COST OF STEAM LEAKS (DOLLARS PER YEAR AT 150 PSIG)

Cost of steam leak when steam generation costs:				
Hole diameter in inches	Leak rate lb/year	$4 per 1000 lb	$6 per 1000 lb	$8 per 1000 lb
1/16	160,000	$640	$960	$1,280
1/8	635,000	$2,540	$3,810	$5,080
1/4	2,500,000	$10,000	$15,000	$20,000
1/2	10,200,000	$40,800	$61,200	$81,600

Check your kiln at least once a week for steam leaks by pressurizing the system and listening and looking for them. In addition, when you change-out charges, be alert to water on the floor, hissing sounds, and new rust marks on the coils. Dark streaks on lumber next to the center coils may also indicate steam leaks. Leaks in drain and condensate lines spray water and are easy to spot. Leaks in feed lines only make noise. You cannot see the steam escaping from them in the hot kiln, and these leaks must be tracked by sound. I was once so close to a large leak that the noise was unbearable, yet I could not see the leak. Only after I passed my hand quickly over it was I able to pinpoint it. Be careful when doing this! If it is not convenient to repair small leaks when found, mark them for later attention. Large leaks demand immediate action (figure 9.16).

First thing in the morning, when you arrive at work, take a close look at your control system, particularly the charts, to see that the kiln ran as expected through the night. Any inability to maintain set points must be investigated immediately and corrected.

If the dry bulb tends to override set point toward the end of the charge, your valves are bleeding through. On old pneumatic systems, this is an indication of seat wear and requires valve replacement or rebuilding for correction. In new pneumatic systems, the valve may simply not be seating. However, remember what I said earlier about valve actuators. If you have a valve system that uses 3–15 psig valve actuators, they may never seat against normal steam header pressures. For this you need 6–30 psig actuators. To adjust the seating on a pneumatic valve, listen to it using a mechanic's stethoscope touched to the valve body on the downstream side. These stethoscopes are available from most auto parts stores and are the same ones I use for listening to traps. A bleeding valve makes a hissing sound. Using a large open-end

Figure 9.15 Testing a trap by pouring water on it may identify a trap that has seized shut, but it will not alert you to the possibility that it is blowing through.

or Channellock™ wrench, turn the bolt on the valve stem to seat the valve. You should see the stem move down as hissing slows and stops. If it does not, then (1) there may be trash in the throat of the valve; (2) the valve seat is worn and must be repaired; or (3) the valve will not close completely against full steam header pressure.

Figure 9.16 Steam leaks must be repaired as soon as possible after discovery, whether in feed or drain lines, coils, or traps. In addition to upsetting the drying balance in the dry kiln, they erode concrete and destroy steel components.

Keep extra valve diaphragms and valve stem seals on hand in case of failure. Diaphragm and seal failures are abrupt and unpredictable and appear as loss or diminishing of kiln temperature, although temperature loss can also be caused by trap failure and the loss of steam pressure.

To track the problem, first make sure you have full steam pressure. Open the valve to 100% through the control system and observe its operation.

- If the control valve is fully open, suspect the trap or a blockage in the condensate system.
- If the valve is closed, then disconnect the valve air line from the controller and place your finger over the controller outlet. If the gauge for that air line does not go to full pressure, then the problem is in the controller.
- If it does go to full pressure, reconnect the air line, and disconnect it at the valve. Place your finger over the end of the air line. If the gauge does not go to full pressure, the problem is in the air line.
- If it does go to full pressure, the problem is in the diaphragm or in the stem seal.
- Place your finger over the vent on the top of the valve. If the gauge goes to full pressure, the problem is a blown diaphragm. If it does not, the problem is a leaking stem seal.

NOTE: Replacing stem seals is a laborious process. I have had some success stopping these leaks by injecting Perma-Seal™ through the air line fitting. It is worth a shot, anyway.

Establish a preventive maintenance program for your steam traps. I used to recommend replacing trap kits routinely once a year on a rotating basis. Many operations may be able to go several years between changing kits. One year after a new kiln has started, pull down a trap and inspect it. Look for wear. As a trap wears, its orifice enlarges, and the small hemispheric ball that closes the trap takes on the appearance of a nipple. As it enlarges, it will have a tendency to seize shut. If the wear is not excessive, run another year and pull down another trap. Continue to do this until you have established how long your traps will run between kit changes and then rebuild all of your traps on that rotating basis. If you repair a few traps per month, you will spread out the cost and downtime, reducing the magnitude of this endeavor. PM programs head off trap problems and allow you to operate at maximum efficiency and energy savings.

If you must replace a trap, there is an easy way to break the unions loose for removal. Using a ball peen hammer, smack at least two flats of the union sharply. You should then be able to unscrew it easily with your pipe wrench. After repair, coat the threads of the unions with Never-Seize™ before tightening. You will find that you can tighten them much tighter to seal leaks, and they will be easier to remove next time.

The buckets in inverted bucket traps will occasionally wear out and have the appearance of having been attacked with a can opener, leaving the body of the bucket in the bottom of the trap. Obviously, the bucket must be replaced. When purchasing new buckets for an existing kiln, or new traps for that matter, specify that they be sent with *large vents*. Otherwise, you will get standard vents. This vent in the top of the bucket is necessary for the removal of noncondensables (air, carbon dioxide, and oxygen) from the system. The presence of noncondensables reduce the heat transfer of the steam by reducing its temperature as well as contributing to internal corrosion of the kiln heating coils. The large vent is required since the kiln functions in a batch process and significant amounts of air can be drawn into the coils between drying cycles. Noncondensables may also be generated in the boiler during steam generation. At any rate, they must be removed, and this vent is the primary reason for the use of inverted bucket traps. If you get standard vents by mistake, the large vent is simply one half of the orifice size and can be easily drilled out.

Once a month take a small bucket into the kiln and discharge the blowdowns ahead of the traps into it. If there are fine black particles in the bottom of the bucket, use a magnet to see if they are attracted to it. Keep an eye on this from month to month. The fine black magnetic particles are the insides of your coils and lines being discharged toward the traps and are the result of carbonic acid generation in the coils. Over time, multiple pinholes develop in the coils, and the drain lines start to clog like arteries. Discuss this with your water treatment company. If it continues, you are witnessing the early stages of the total collapse of your steam heating system. I have run into this situation in Montana, Mississippi, South Carolina, and Mexico, and it indicates a problem with the water treatment at the boiler. At the above locations this was a problem with M-alkalinity. In extreme cases, the installation of a dealkalizer is recommended to correct it.

REFERENCES

1. Armstrong Machine Works. *Steam Conservation Guidelines for Condensate Drainage.* Three Rivers, MI: Armstrong Machine Works, 1976.
2. Culpepper, L. G. "Application of Computerized Controls to Improve the Lumber Drying Operation." Seattle, WA: IUFRO Meeting, 1989.
3. Hammond, H. "Principles of Operation to Be Considered in Selecting Steam Traps." *Plant Engineering* (July 1985): 70-73.
4. Makansi, J. *Managing Steam.* New Jersey: Leslie Company, 1985.
5. O'Dell, L. R. "Steam Traps—Basic Facts and Energy Saving Tips." *Heatinq/Piping/Air Conditioning* (May 1977): 53–58.

Chapter Ten

Direct-Fired Dry Kilns

Education is when you read the fine print.
Experience is what you get when you don't.
Pete Seeger

Operation

Direct-fired dry kilns are heated by blowing hot combustion gases directly into the kiln environment (figure 10.1). These gases are generated by burning oil, gas, or wood waste. Historically, high fossil-fuel costs dictated that direct-fired kilns were built to burn wood waste except at mills where natural gas was available and where product lines required bright lumber for end users. Until the late 1980s most wood waste burners burned hammermilled planer shavings for fuel, and the lumber produced was characteristically gray to black due to the deposition of ash on its surface. This precluded its use on kiln dried after treatment (KDAT) lumber or lumber that was earmarked for rough export. With the advent of burners that utilize green sawdust for fuel, the ash problem has been significantly reduced.

Shavings burners require a small, uniform particle size for efficient combustion (figure 10.2). Shavings are blown directly from the planer mill into a hammermill, which sizes the fuel and blows or conveys it to storage bins. This fuel is then metered into the burner, where it mixes in a cyclonic action with air as it burns. The products of combustion (heat, water, carbon dioxide, and ash) enter an air circulation system for distribution into the kiln. The ash deposited on the lumber is completely removed during planing, and as such these kilns have found wide use for dimension lumber. These kilns are generally not suitable for mills marketing rough dry lumber. The control of ash particles at the planing mill under increasingly stringent OSHA mandates is also an area for concern.

Figure 10.1 Direct-fired kiln utilizing hammermilled planer shavings for fuel.

Figure 10.2 Close-up of shavings suspension burner.

Sawdust burners burn green sawdust directly on a grate, not unlike a boiler firebox. Burning green sawdust does not recycle ash through the system as happens when burning shavings, and since little fuel becomes suspended in the airstream, these burners generate less ash (figure 10.3). Lumber from these kilns is brighter, although this varies between manufacturers and the individual level of maintenance. If overfired, they will also produce ash. In general, though, the surface color of the lumber dried in these kilns approaches the color of lumber dried in steam-fired kilns.

In areas where natural or LP gas is available, you will find direct-fired kilns built to use this fuel. The color is indistinguishable from lumber from

Figure 10.3 Direct-fired kiln burning green sawdust. The truck in the foreground collects and hauls away ash from the burner.

steam kilns, which permits its use on higher grade lumber and for KDAT. Like other direct-fired kilns, there is no provision for a steam spray system unless it is added as a separate system. These kilns are not recommended for lumber going to end users who will remanufacture your lumber unless a spray system is provided for equalization and conditioning.

Direct-fired kilns firing wood waste are somewhat less expensive to purchase than steam kilns, although the more kilns you plan to build, the more competitive steam becomes. Those firing natural gas are much less expensive than steam kilns and are a logical choice for certain mills.

My first book was highly critical of direct-fired kilns for the following reasons:

1. Ash buildup on the lumber is a problem for many.
2. Grade recovery runs somewhat below what was possible in a steam kiln.
3. These kilns are more prone to fires, which occur most often toward the end of the charge, when the lumber is driest.
4. Shavings and sawdust burners require more electrical horsepower for their operation than a comparably sized steam kiln, including boiler.

Concerns with ash have been addressed with the newer green sawdust and gas-fired kilns. Grade expectations have dramatically improved. Indeed, for many applications a well-maintained and managed direct-fired kiln can be on par with steam. The fire issue remains with shavings burners but has been significantly minimized with the sawdust burners. Gas kilns are generally very safe.

The key to satisfactory performance of these kilns is proper heat balancing and burner maintenance. Direct-fired kilns are characteristic in their variations in air velocities in different areas of the kiln owing to the eccentric nature of the heat distribution system. Keep direct-fired kilns short, certainly no longer than 68 feet, to minimize and manage this variation.

Beyond the choice of burners, the design of these kilns is similar between manufacturers. For the purposes of this discussion I will treat all burners simply as heat sources in discussing the general operation and maintenance of these kilns.

Heat Distribution Systems

Direct-fired kilns are equipped with a ductwork system that distributes heated air from the burner into the kiln (figure 10.4). Heated products of combustion are discharged into the blend chamber where they are mixed with relatively cooler air returning from the kiln through the return air duct. A large centrifugal blower circulates air from the kiln and through the blend chamber, blowing it into the crossover duct. Some manufacturers place this fan ahead of the blend chamber, forcing the burner to work against its output pressure. Both the distribution and return air systems are sources of imbalances in heat and airflow.

The crossover and main distribution duct may be underground in older systems, although this design is not seen very often today. Heat losses through the ground are high and distribution efficiency poor. The smaller ducts branching up from this are called risers, since they rise from the ground to the upper fan deck area between the two sets of tracks.

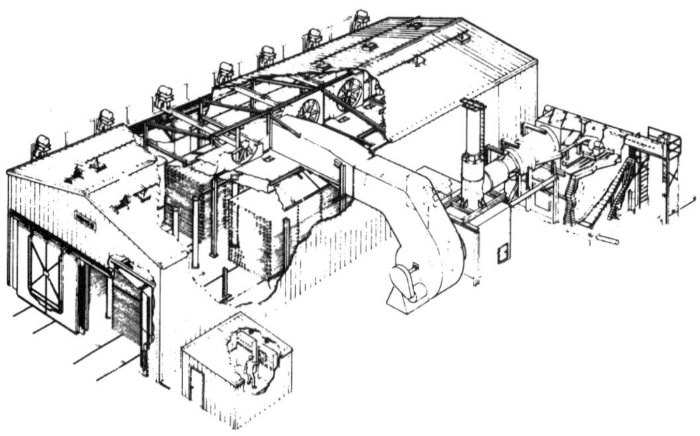

Figure 10.4 Schematic drawing of a direct-fired kiln equipped with a burner designed to burn wet sawdust. Illustrates burner, blend chamber, return air duct, recirculation blower, supply and distribution duct, and downcomers. Drawing courtesy of HEMCO.

Most crossover ducts run overhead and connect to the main distribution duct built as an integral part of the fan deck. This allows delivery of heat to a central location for uniform distribution between the tracks and overhead. Avoid any design that places a single main distribution duct to one side of the kiln within one of the plenum chambers. Serious imbalances result from this design. From the main distribution duct, heated air flows in two directions. Downcomers extend down to the ground between the two sets of tracks and serve as the main heat distribution path (figure 10.5). They are square or diamond shaped in cross section, slotted on two sides along their length, and fitted with dampers on their inlets. Additional openings are cut into the top or sides of the main duct, which introduces air into the overhead fan deck area. These openings are equipped with slide gates or other airflow adjusting mechanisms. Sometimes openings are simply cut into the duct with a torch.

Figure 10.5 Downcomer extending from underside of main distribution duct between the tracks of a double track kiln. Note installation of baffle at top to cut down on heat in this area.

Heat Balancing a Direct-Fired Kiln

The real trick to balancing moisture distributions in these kilns is balancing heat delivery. This is a difficult and time-consuming process and involves balancing heat delivery from the downcomers along the kiln length. We then balance the downcomers with the overhead duct openings. Success is often illusive, since any adjustment to any area changes the balance of the entire kiln. Close a damper or slide gate here, and the air pushes out somewhere else. Heat balancing is truly an art.

If necessary, wash the kiln to remove any ash buildup before beginning. Pay close attention to the slots in the downcomers—they must be completely free of ash. Ash buildup is a serious and continuing problem in direct-fired kilns that burn hammermilled shavings. Ash gradually accumulates in these kilns, restricting and rerouting airflow through the distribution system, upsetting heat balance. Indeed, you may find it necessary to wash frequently through balancing procedures lasting more than a few charges.

Downcomer Washing Procedure

Using a fire hose, direct the water stream into the downcomer slots from top to bottom. Make certain all ash is removed from both slots. Wash all the downcomers in the line. Since you can wash only one side of the slot from any given track, switch to the other track and wash from the other side. Only then will you remove all of the ash accumulation. Inspect the fan deck to see if the openings here require washing. Make washing a regular part of your maintenance schedule.

Inspect the downcomers for damage. Look for broken or loose nuts and spacers. If downcomers are deteriorating where they contact the floor, repair them before proceeding. Check the downcomer dampers. These dampers are easier to adjust if they have handles on them, but all can be moved with a wrench if needed. Installing handles aligned with the damper blades makes it easier to determine damper position from the ground. Make certain the dampers have lock nuts or U-bolts to keep them in position, otherwise you will want to tack weld them into position when you are satisfied with the balance. Coe direct-fired kilns are usually built with two sets of dampers in the downcomers, one at the top and one at the midpoint of their length. Open the midpoint downcomer 100% and lock it in place or remove it altogether. It serves no purpose.

Turn on the recirculation and combustion air blowers. Inspect the downcomers again. Also inspect above and below the overhead distribution duct. Bolts often fall out of the duct and are unobserved unless the duct is pressurized. Repair any gaps that allow the uncontrolled escape of heated air into the kiln.

You are now ready to begin balancing the downcomers.

Downcomer Balancing Procedure

Tools required: screwdriver; Allen or open-end wrench to loosen damper set screws; Channellock or pipe wrench if the dampers do not have handles; lumber crayon; ladder; and an airflow meter. A helper speeds this process since he can adjust the dampers at the top while you monitor the air velocity meter from the ground.

1. Number each downcomer with your crayon so you know which downcomer you are working on at all times.
2. With the recirculation blowers on, begin measuring the air velocity out of both downcomer slots at chest height, orienting the outlet of the airflow meter towards the return air duct. Write these readings directly on the downcomer as you work your way down the kiln length. These numbers will provide handy reference data later.
3. When finished, record the data from each downcomer on your data sheet. The average of both downcomer slots is the average flow value for that downcomer. These are the values we will use to adjust flow.
4. Examine your data to determine where your highs and lows are and develop a plan of action.
   - Adjust low downcomers up first. You may find that these are already full open and that there is no adjustment left. We will deal with this problem in a moment.
   - Next, adjust high downcomers down. Many of these values may be lower than when you first checked if you opened other dampers.
   - Remember, we are trying to reduce the variation between the downcomers to obtain more even heat distribution.
5. Make another pass through all the downcomers as in steps 2 and 3.

Now comes decision time. If there are no seriously low or high downcomers and you are satisfied with the variation, you may stop. If you believe that you can get them closer, then keep repeating steps 2 through 5 until you are satisfied. I like to get them within a range of around 100 fpm.

> NOTE: On newer kilns you may find that your airflow meter is pegged out due to the high volume of air the newer systems generate. No problem. Simply duct tape over a portion of the opening to cut down on the inlet size of your meter. We are looking for relative differences between the ducts and not absolute values anyway.

Table 10.1 shows sample downcomer readings from a balancing procedure. Downcomers nearest the connection between the supply duct and the distribution duct usually have the highest velocities (numbers 5 and 6). Closing these pushes additional air out toward the ends. There will always be one or two downcomers that never seem to get enough air, even when fully opened (number 7). This is apparently caused by air rushing past the top of their openings, unable to make the downward turn. In newer kilns, I have found that this is likely due to the proximity of downcomer openings to support columns within the duct. You may be able to correct this tendency to overshoot by installing 3 × 3 angle-iron scoops within the main distribution duct to direct air into the deficient downcomer and help build up static pressure. I agree this is not easy. The only reason I even attempted this was that my Daddy was short. It generally works, though, and is why I was able to balance this kiln so close. These crossover and main distribution ducts should be designed and built with access for inspection, anyway. Occasionally you will find that installing the scoops has a negligible impact on directing flow into the downcomer. Here you may have success closing selected slide gates on the top of the main distribution duct in the vicinity of the offending downcomer to increase static pressure within the duct.

TABLE 10.1 SAMPLE DOWNCOMER READINGS IN FEET PER MINUTE

Downcomer		Before adjustment Velocity		After adjustment Velocity		
Number-slot		**Reading**	**Average**	**Reading**	**Average**	**Change**
1	L	1240	1320	1670	1685	+365
	R	1400		1700		
2	L	1670	1740	1670	1670	-70
	R	1810		1670		
3	L	1260	1315	1660	1655	+ 340
	R	1370		1650		
4	L	1300	1330	1680	1660	+ 330
	R	1330		1660		
5	L	2320	2310	1740	1690	+ 620
	R	2300		1640		
6	L	2270	2270	1660	1650	- 620
	R	2270		1640		
7	L	840	935	1660	1645	+710
	R	1030		1630		
8	L	2020	2165	1690	1675	- 490
	R	2310		1660		
9	L	1000	1080	1690	1660	+ 580
	R	1160		1630		
Average		1607		1666		
Maximum		2310		1690		
Minimum		935		1645		
Range		1375		45		

Do not discount the difficulty of achieving good balance. Although it appears that we balanced this kiln in one simple step, there were actually three intermediate balancing passes and readings made in this kiln. The entire process consumed an entire day. You should expect more variation on longer kilns.

At one time I believed it was important that the downcomer opening slots be tapered from top to bottom, with the slot widest at the bottom. I reasoned that the air pressure must be highest at the top of the downcomer and more air must escape there. This would help explain the warped lumber at the tops of units. The next time you check the inside of your kiln with the circulation blower on, take a piece of ribbon and hold it in the airflow from a downcomer slot. Move it up and down along its length and observe how the ribbon follows the flow. You will discover, as I did, that air does not blow straight out from these openings, perpendicular to the length of the duct. It tends to flow downward toward the floor. Tapering the slot opening seems less important when viewed in this light.

Overhead Opening Balancing Procedure

Since there are so many openings in the fan deck level, balancing these is a real challenge. Not only are these openings affected by the openings in the downcomers below them, fan direction also influences how much air is drawn from any one opening depending on which plenum is pressurized. Also, anything you do here will affect the balance of the downcomers. Although I have known individuals who claimed they could adjust these by sight, mere mortals must rely on trial and error. You make an adjustment, run a charge, make another adjustment, and so on. How do you know how well you are doing? Most direct-fired kilns are installed with high-temperature limit sensors in the roof. These sensors mount on both sides of the fans at regular intervals along the kiln length. Although a little hard to read with accuracy, these sensors often supply the only information available to aid balancing the overhead fan deck openings. Newer kilns have multiple heat sensors installed along the kiln length within the plenums. However, no temperature readings are reliable unless your kiln is fully loaded and properly baffled. Openings in the lumber that permit circumvention of air contribute to hot spots and erroneous readings.

Keep in mind that higher-pressure areas deliver more air through a given opening. As such, openings tend to be smallest near the supply duct junction and open gradually toward each end of the kiln, with one exception. At the supply duct junction they need to be open a bit more to overcome the reduced openings in that area. Simple, right? Now you have an appreciation for the art of balancing these kilns.

In steam-heated kilns we try to design our heating surface with approximately 60% between the tracks and 40% overhead. I have heard discussions of the merits of this percentage split versus one which is evenly split. These discussions are all well and good, but at no time has anyone been able to demonstrate to me how they determine when the heat distribution is split to any percentage.

Checking your progress in any heat balancing endeavor is much easier with multiple heat sensors positioned at strategic locations within the kiln. Permanent thermocouple networks designed by the kiln manufacturers or made in-house are very useful. These systems allow you to monitor the kiln for changes that will occur over time. They are also easy to use. A good system prints out temperature data at regular intervals and presents it in an easy-to-understand format. It is frustrating and time consuming to have to pull data out of column after column of numbers to make any sense of it. Also, the easier it is to use, the greater the likelihood it will be used on a continuing basis. One note of caution: If your sensor network uses two wire RTDs instead of thermocouples for temperature sensors, be certain they are in calibration. I once spent a week trying to tune a hot spot out of a kiln only to discover that the hot spot was a single sensor 20°F out of calibration.

For temporary measurement, rent a multipoint recorder from an instrument house. Inexpensive 22-gauge thermocouple wire with nylon or polyurethane insulation is available off the shelf and may be discarded after use. For extended testing, or for use in several kilns, consider teflon-insulated wire. You need sufficient wire to map the heat distribution down both plenums and between the tracks of your kiln. Make your thermocouples up by attaching one end of the wire to the recorder, pulling the wire into the kiln, and clipping it to length. Strip the leads back about 1 inch and twist them together. Slip a crimp connection over it and crimp. Trim any exposed ends. Dip the end in RTV silicone sealant or slip a piece of heat shrink filled with silicone over the end and shrink it.

Use your thermocouple network to determine where your hot spots are, making adjustments to dampers between charges. Over time you will narrow the heat distributions in the kiln. This reduces moisture variability in the lumber. Do not allow ash buildup to fool you into making erroneous changes. You will have to wash down the ash frequently during balancing in those kilns that experience this problem.

Balancing Return Air

So far we have discussed kiln balancing in terms of balancing heat delivery. If you run a few charges with a thermocouple network, you will soon discover that hot and cold spots seem to move about the kiln with fan reversals. A contributing factor in this is the action of the return air duct.

The return air duct is necessary to return kiln air to the blend chamber for reheating. Almost all of the air distributed by the recirculation blower is returned through this duct. The difference is in the additional air added to the system by the combustion air blower. That air volume escapes the kiln through the vents and cracks in its walls. It also depresses humidity levels. Still, a tremendous volume of air is recirculated.

Most kiln designs use a single return duct mounted off-center on one long wall (figure 10.6). Natural airflow patterns are radically altered, since virtually all kiln air is migrating toward this duct. The duct has a boosting effect on air velocity when air circulates through the lumber toward it. It reduces air velocity in its vicinity when fans run in reverse.

Return air duct openings must be shielded to reduce heat radiation onto the lumber in their vicinity. These shields stand straight up or are angled against the wall. There does not seem to be any particular advantage to either design. However, air must be able to pass freely around its entire perimeter. A few manufacturers cover this opening with expanded steel mesh to prevent hard hats and kiln operators from being sucked into the blend chamber. Expanded steel mesh clogs rapidly with ash, reducing the efficiency of the recirculating blower. Replace this steel mesh with 6 × 6-inch weld wire.

The solution to the uneven air migration problem is to split the single return duct into several ducts, spaced at regular intervals down the length of the kiln (figure 10.7). This spreads out the withdrawal of air and minimizes the impact of the single large duct. A few mills experimented with splitting this duct by running the connecting ductwork within the kiln plenum chamber. These designs did not work well since they reduced the plenum chamber volume at the bottom of our carts, the worst possible place. We want the plenum chamber to be as wide as possible top to bottom to minimize top to bottom airflow variation. Otherwise, airflow is reduced to the bottom lumber layers, increasing our problems.

In 1983, I designed a split return duct that was installed on the outside of a kiln in Meldrim, Georgia. We mounted this duct between the kiln and the control room with openings cut into the kiln side walls. The improvement in airflow uniformity was dramatic. Since it was no longer necessary to overdry portions of the kiln to get other areas dry, drying time reduced and grade improved. Still, the split return duct does not address the velocity drop that accompanies changes in fan direction. Most kilns experience velocity drops of 100 fpm or more between fan directions, which shifts the highest moisture content boards of our carts to one side. This is best overcome by running the fans longer in one fan direction to compensate for the velocity differences

Figure 10.6 Single return air duct equipped with shield. Note installation of plenum baffle.

and moving the highest moisture content boards to the center where they should be.

A few attempts have been made to install return ducts on both sides of the kilns. Dampers switch between the two ducts with fan direction. I have no information on the success of these efforts and am unaware of any in present operation. In theory, placing a split return duct on both sides of the kiln would give the most uniform drying for these kilns, although the cost of doing this may well outweigh the benefit. By the time we install all of the equipment, bells, and whistles required to overcome the known imbalances in direct-fired kilns, we could likely have purchased a steam-fired system that does not share these problems.

Figure 10.7 Exterior view of split return duct.

Handling Ash Deposits

Ash buildup is a minimal problem with the newer sawdust burners and has never been a concern with gas-fired. Shavings burners are another story. Ash buildup in these kilns is perhaps the most common complaint about direct-fired lumber. The smoky gray appearance it imparts to the lumber surface is objectionable to customers purchasing rough stock. This includes customers in the export markets and treating plants purchasing dry, rough timbers. Since most sawmills realize this, these kilns are built only at mills that plan to dress all of their production.

This ash causes problems at the planing mill since it shakes off the lumber as it is unstacked, conveyed, and fed to the planer. The health impact on planing mill workers is gaining increasing attention. New OSHA standards regulating airborne dust requires its elimination from the workplace, and personal protection devices are not acceptable under the new rules. This places a tremendous burden on the employer to remove or control this dust before worker exposure. Anyone who has visited a planing mill that dresses this lumber is immediately struck by the fine layer of dark dust on everything. Mill managers recognize this problem and have attempted to deal with it in a variety of ways.

The most common method involves spraying a fine water mist on the lumber as it is unstacked to prevent the dust from becoming airborne. This works fairly well, provided an effort is made to capture and remove the small amount of dust that shakes off the underside of the lumber. The working area around the unstacker still tends to become very dirty. Water runoff must be collected, settled, and recycled to avoid the discharge restrictions placed on process water in industrial plants. Water absorption by the lumber does not tend to be a problem because it remains on the surface for a very short time, and the surface is dressed off anyway.

A few mills have placed hoods directly over the unstacker in an attempt to suck dust off the lumber and out of the planing mill. In general, this only collects the dust that actually becomes airborne in the vicinity of the unstacker. General working conditions tend to be dusty all the way to the planer. A variation on this is to enclose the entire unstacker, infeed conveyors, and planer. This significantly reduces the dust problem outside these areas and appears to be the most efficient method for addressing the problem.

Experiments have been made using a rotary brush system ahead of the planer. This cleans up the shavings but does not address the airborne dust problem that occurs ahead of it. In actuality, this method has some merit when used from the beginning and on a continuous basis. Direct-fired lumber tends to become darker the longer a kiln is in operation, particularly in water spray systems. Since planer shavings are used as fuel, ash tends to recycle and build up in the system. Brushing ash from the lumber before it is planed reduces this buildup. New kilns always produce bright lumber with little accumulated ash when started up for the first time.

One mill experimented with a blower system mounted at the kiln doors designed to blow the ash off the lumber as it was pushed from the kiln. This approach was apparently unsuccessful. Had it worked we would likely be seeing more of these in operation today.

Manufacturers of these burners have experimented with different methods to reduce ash generation. Their efforts include slotting the outside of the burner at the choke collar to sift out incompletely burned fuel particles, and extending the length of the blend chamber to increase its volume and allow ash to settle out. All have met with some, although far from complete, success. The advent of the sloped grate green sawdust burner has made serious inroads into this market with good reason. It burns less expensive fuel and does not share the ash problem.

Direct-Fired Control Systems

Almost all direct-fired kilns are set up with single-zone control. A few have been built with two zones, in effect dividing the kiln lengthwise into two single-zone kilns. These kilns invariably experience control problems in regulating and balancing heat distribution to the two ends.

With single-zones, most direct-fired kilns use traditional pneumatic controllers. These may be RTD or the older gas-filled systems. Newer kilns use modern digital controllers with thermocouples as sensors. Computerized controls have now come into being for these kilns and serve to monitor and control kiln schedules as well as the controls for the burners. They are very efficient and work extremely well.

Since most direct-fired kilns monitor supply and return air duct temperatures, their temperature difference can be related to lumber dryness. This works similar to the TDAL (temperature drop across the load) method for determining when to stop the drying process. Fan direction is important when using this method to pull the kiln, and you should always start your kiln cycle with the fans blowing in the same direction to get consistent, reliable data. You will get to set point faster by starting cycles with the fans blowing air through the lumber in the direction toward the return air duct.

I have seen several ideas for zoning these kilns into more than one or two control zones. A few would attempt to regulate heat top to bottom within the kiln similar to multizoned steam kilns. None have been implemented. Be very cautious in considering these proposals—there are some very difficult technical problems. Until now, designs have been limited by the service conditions actuators must endure in these kilns. I expect this limitation will be overcome by good kiln design or the availability of actuators designed for severe service. The real problem is more difficult. Since all control heat is delivered by a set air volume carried through interconnected ductwork, any change in air/heat delivery in one area results in additional air volume elsewhere. Close a damper here, and its air pops out somewhere else with unpredictable results. Balancing control to the different areas of the kiln will be complicated, risky, and expensive. These systems will doubtless approach steam kilns in cost, without their reliable maintenance and quality track record.

Proven Suggestions to Improve the Operation of Your Direct-Fired Kiln

Here are a few proven suggestions to help your system run more efficiently:

For Wood Waste Burners

1. Clean the kiln on a regular basis, paying close attention to ash buildup on duct outlets and inlets. Extremely dusty systems require washing once a week. Clogged delivery systems extend your drying time and increase lumber moisture variability.
2. Scrape the ash buildup from your fans and recirculation blower at least once a month in those kilns that experience this problem. Allowed to accumulate, chunks break loose, unbalancing the blades and causing premature failure. If you are losing a lot of fans, this may be the reason.
3. While you change out charges, leave the burner on when running wood waste systems. This extends the life of burner refractory and shortens kiln heat-up time. Gas-fired kilns are always shut down between charges.

For All Direct-Fired Kilns

1. You will get to set point faster by starting cycles with the fans blowing air through the lumber toward the return air duct.
2. I maintain that top-layer warp in these kilns is due to a lack of control over the top to bottom heat distributions in the kiln. However, there are a couple of other ideas that have been tried that have helped this problem.
   - At least one manufacturer has started adding a sheet metal baffle approximately 3 feet wide extending the length of the kiln. This baffle is connected to the underside of the main distribution duct and passes along the side of the downcomers between the tracks. This seems to affect airflow through the top few feet of the stacks and reduce warp there.
   - The main distribution duct has a surface temperature approaching 500°F in many kilns. I believe that the radiation off this duct is a contributing factor in top-layer warp. Individual mills have installed a light, retractable plate within their kilns, which lays on the top layers of the units, presumably to hold top boards in place. The top layers do lay much flatter than in kilns that lack them. I suspect the plates' success comes from blocking this radiation and wonder if we might attain similar results at far less expense by simply insulating the underside of the main distribution duct.

3. There has been significant success increasing the output of the recirculating blower by increasing its RPM. This reportedly speeds heat-up and results in more uniform heat distribution. Indeed, the most recent kilns I have seen have phenomenal downcomer velocities, and I can report that the drying quality is outstanding, although doubtless for a number of reasons. Before you speed up the blower, discuss this with the blower manufacturer and follow his recommendations.
4. I have recently encountered a number of gas-fired, direct-fired kilns built in the Midwest that typically run a very low distribution duct temperature. These kilns are typically set up to run KDAT lumber or other treated building materials and are often kept to a temperature of 200°F or less owing to a misinterpretation of the rules governing the maximum temperature you can dry these materials at. You can significantly improve the performance of these kilns by raising their duct temperatures 25°F to 50°F without fear of violating the temperature rules.

Chapter Eleven

Computerized Lumber Drying Controls

You only live once, but if you work it right, once is enough.

Joe E. Lewis

As a lumber inspector for SPIB during the latter part of the 1970s, I visited many lumber mills throughout the South. I was repeatedly struck by the disregard many of them felt toward their dry kilns. Indeed, this was the impetus for me to make dry kilns my career. Too many mills viewed their dry kilns as a necessary evil, a black box into which they placed green lumber and removed dry. Poor grade recovery was accepted. Few mills appreciated the real impact good drying would have on their sawmill. After all, the real business was sawing lumber and production was king. Chip-n-Saws and double arbor gangs were exciting additions to the sawmilling trade. All problems could be solved with just a little extra production.

Rising log costs began to get everyone's attention in the early '80s. There was a sudden scramble to reduce saw kerfs and install scanners to reduce waste and improve log breakdown, board edging, and trimming. Gradually, mills began to realize that the true value of their lumber lay in what made it out the back door. What made it out the back door was heavily influenced by what happened to the lumber in the kiln. The kilns, in turn, were heavily influenced by how they were designed, operated, and controlled. In 1982, an evolution in control systems for dry kilns began. That year saw the introduction of computerized, multizone drying controls on two high-temperature kilns at Union Camp Corporation's Franklin, Virginia, pine sawmill (figures 11.1 and 11.2). The design of these kilns was based on my research and development over the previous two years. The benefit of these systems has since extended to all lumber species, since the

Figure 11.1 The first computerized, multizoned dry kilns were installed at the Union Camp mill located in Franklin, Virginia. Photo courtesy of International Paper.

Figure 11.2 Visual comparison between green lumber and dry lumber from a computerized multizoned dry kiln. Aside from the color and height differences, there is little to distinguish the two tracks. Top layers lay flat.

need for good drying control is universal. These controls are now approaching two decades of use and have proven their benefits in grade, energy, and drying time reductions time and again. You can no longer afford to run kilns without computerized controls.

Computerized Controls

The original computerized system used a PLC-type process controller, which programmed in ladder relay logic. This controller printed process data at regular intervals and would alarm motor and sensor failures. Interpreting the printed information required a thorough knowledge and understanding of the process and equipment. Still, this system was a resounding success for Union Camp. Overall grade recovery increased from an average 70.7% #2&Btr to 81.9%. Energy costs reduced 43% for steam and 10% for electrical power. Total savings including grade, energy, labor, and maintenance resulted in a 1.2-year payback on investment. In 1988 this computer system was replaced with advanced computer controls manufactured by a major kiln supplier, but still patterned after the original. The percentage of #2&Btr increased an additional 2%, with a slight reduction in drying time.

Following is a listing of the demonstrated benefits of computer controls for dry kilns:

- *Grade Improvement* Conversion of existing kilns to computerized control results in a modest 2% to 5% improvement in grade. Conversion to computerized multizone control has seen improvements varying from 4% to 12%. Individual results depend on the type of drying done before conversion, whether the kiln was converted using the same drying methods, or if controls alone are replaced on an existing system.
- *Moisture Distributions* Moisture distributions narrow in kilns converted to computerized, multizone controls. Simply converting to computer controls without multizoning is not known to have any impact on moisture distributions, since you have not really changed the drying characteristics of the kiln.
- *Drying Time* If you are converting existing kilns to computerized multizone controls, expect to see drying times reduced approximately 10% for high-temperature southern pine. Longer-drying species using conventional schedules have seen reductions approaching 40% but generally enjoy reductions of approximately 25%. When controls alone are substituted, you may see little change in drying time. Many mills have discovered that by upgrading the controls and heating systems they may not have to build additional kilns to increase drying production. Others find they can shut down inefficient kilns with the increase.

- *Energy* Energy savings through conversion to computerized, multizone dry kilns range from 17% to 43%.

  NOTE: Your conversion results will depend entirely on the technology of your individual supplier.

Talk to, and visit, customers of potential suppliers of this equipment before committing to a decision. If you have any doubts, require a reasonable performance guarantee. But remember, the computer does not replace good dry kiln practices or good dry kiln personnel. The greater the understanding and ability of the people who use it, the greater the potential benefits. The computer does not think, but only reacts to situations as it has been programmed. It cannot compensate for things you forget or fail to do. Do not expect the computer to do everything. Also, there is no universal drying schedule suited to every kiln. Each kiln is unique and must be treated as such. It is the responsibility of the operator to discover what works best for his situation.

The best source of drying schedules remains the *Dry Kiln Operator's Manual* published by the Department of Agriculture. For computerized control systems these schedules should be treated as suggested guidelines only. These schedules have simply not kept up with the evolving technology.

All of the schedules in the *Dry Kiln Operator's Manual* were developed using the old-style pneumatic recorder controllers. These controllers were the reliable workhorse of the drying industry up until two decades ago. They are now being replaced with digital equipment. When I first started in this business I ran 11 track kilns which had these controls. In the morning when I first arrived at work I picked up a small tackle box holding the tools and parts I needed to service and maintain these controls, along with sufficient tools to affect minor repairs. I carried this with me as I made my rounds, checking charts, adjusting set points, and generally assuring that the kilns were running as expected. I did this again in the afternoon before leaving for home. The schedules reflected my morning and afternoon routines and were designed to be adjusted every 24 hours. This is how they had been run for decades, and they would doubtless still be run in this manner, had we not demolished them to make way for the two new kilns that replaced them. The point is, the schedules were set up for the convenience of the operator as are all of the schedules in the *Dry Kiln Operator's Manual*. Even a cursory examination of the schedules in this manual will satisfy you that all temperature changes occur at 24-hour or perhaps 12-hour intervals. This is not coincidental but recognizes the fact that a kiln operator cannot monitor even a single kiln on a continuous basis. He has other responsibilities as well.

Even with the first system we recognized that we had potential capabilities far beyond the controls in existence and took modest advantage of them. We ramped set points, controlled on exiting air, multizoned, and shut down on temperature drop across the load. At that time, these were innovative ideas. Since then, computerized controls have opened up a whole new world in drying technology. We are trying and using schedule styles and control schemes that were never even imagined until a few years ago.

In reality, it is doubtful that published schedules will ever be able to keep pace with technology. The older schedules work well because they are by nature very conservative. The range of kiln designs and situations at the individual mills means that the schedules presented have to run on a wide variety of systems. The developers cannot afford to be responsible for schedules that do not work on the odd piece of equipment.

Today, I use these published schedules as guidelines for how the lumber I am drying reacts to moisture removal. Looking at a schedule for a new species or item and comparing it to a familiar one gives me a clue as to drying rates. The manual will also tell me if a particular species is susceptible to specific drying defects. From this information I build schedules, taking into consideration the capabilities of the specific equipment I am developing the schedules for. I can always dry faster than the published schedules and will reasonably expect 15 to 25% reductions in drying time over published values. Occasionally I can cut it in half; it just depends on the individual situation. However, to get the most out of any dry kiln, I want the capability and flexibility to monitor and control every aspect of the kiln operation. The following equipment options are valuable tools in accomplishing this:

Basic system components:

- Latest computer and operating system with sufficient RAM and disk storage to run the kiln control program as well as spreadsheet and word processing programs. Being able to archive and retrieve each charge history is paramount to the effective management of my kilns.
- Large color monitor.
- Printer with graphics capability.
- Uninterruptable power supply (UPS).
- Modem with remote monitoring and control capabilities.

Control software minimums:

- Screen displays to monitor each section or station in the kiln.
- Graphs of kiln parameters including:
    1. Average dry bulb temperature (entering air and exiting air) and set point.
    2. Wet bulb temperature and set point.

3. Steam pressure.
4. TDAL. You should be able to scale it separately from the rest of the chart so it is easier to read and interpret. Its shape gives very important information.
5. Vent and spray position.
6. Estimated or actual steam demand or average percentage steam valve openings.
7. Fan speed if using a variable speed drive.

- Multiple schedule storage capability. You want storage capacity for one schedule for each item that you dry for each kiln on the system. Kilns are individuals; treat them as such.
- Maintenance screens that permit you to monitor temperature sensors, stroke valves, vents, and sprays, and run fans in both directions and adjust their speeds if a VSD is on the system.
- Alarm screens and software that monitors each system and alerts you to problems such as
  1. Temperature sensors high or low.
  2. Suspected fire.
  3. Power outages at the I/0 panel.
  4. Control zones not heating up.
  5. Wet bulbs out of range.
  6. Vents not opening.
  7. Motor failures.

- Charge history archive capability. We want all of the kiln charts stored on hard disk along with run information that includes
  1. When charge started and ended. How long charge spent in each segment of the schedule: actual drying time, length of equalization and conditioning steps, length of non-drying time (pause or hold) within the run.
  2. Shutdown TDAL, ending wet bulb level.
  3. Number of alarms and their nature.
  4. Hot average moisture content and standard deviation, as well as planer mill average moisture content and standard deviation.
  5. Ability to add notes on problems encountered during the run.
  6. Information on how much lumber was in the charge, as well as the lengths loaded.
  7. The actual schedule the kiln ran under.
  8. Retrieving histories must be easy to accomplish. Having to flip through multiple screens to obtain drying information is very distracting and discourages you from your objective.

- Event archive for each individual kiln and for the system as a whole. This permits you to reconstruct problems by recording their sequence of events.
- Alarm archive to also assist you in determining what happened in the event of a problem.
- The ability to start, stop, hold, or pause a charge at will, knowing the schedule will pick up where it left off when it is resumed.
- You should be able to advance the time into an existing drying schedule without altering time parameters within the schedule. This eases restarting a kiln that was allowed to shut down prematurely.
- Should you make a mistake, you should be able to undo errant or mistaken keystrokes without having to shut down and restart a schedule.

Schedule features:

- The ability to ramp all set points (dry bulb, wet bulb, fan speed) up or down, at any rate, and at any point in the schedule.
- The ability to select entering, exiting, or average air heat control is useful for tuning schedules for specific kilns.
- Set up individual parameters for determining when to shut the system down: time, TDAL, wet bulb reading, or energy use. The ability to select a combination of factors is useful.
- Adjust fan reversal times at various segments of the schedules.
- Separate equalization and conditioning segments that run at the conclusion of the regular schedule and are not a part of it. We would, of course, ramp between set points.

Heat control:

- Full modulated heat control. We want to control the valves throughout the entire range of their operation. Temperatures should hold steady rather than cycling over and under set point. On/off valves are not acceptable.
- We want the option to use entering air, exiting air, or average air heat control as conditions require.
- Multiple zone control capability is a must but dependent on the design of the individual kiln.

Humidity control:

- The option to use conventional, exhaust or demand venting.
- Full independent modulated control of both vent rows.
- Full modulated control of the steam spray valves.

Fan control:

- Standard fan reversal capability with intervals fully selected throughout the schedule.
- Failure alarm if vents fail to open prior to reversal.

- Stage start capability if there is no variable speed drive (VSD). We want to be able to stage start the fans on startup or run half the fans for an initial heat up prior to bringing them all on at our option.
- We want to be able to minimize times during reversals by adjusting the dwell time during which the fans coast to a stop.
- With a VSD capability:
  1. Minimize reversal duration by using the VSD as a brake.
  2. Monitor amperage on fan system to run fans to a maximum amp load. If not,
  3. Adjust fan speeds at each segment of the schedule, ramping between segments.
  4. When do we get the capability to exceed 60 Hz?

Mounting Temperature Sensors

The temperature sensor system for your kiln is the window on your drying world. If temperature sensors are not properly positioned to give you the correct information by which to control the process, then you are drying blind. For all systems, the temperature sensors must be mounted in an active air stream and not tucked off in a corner somewhere. This can be a particular problem with package kilns, especially those of European manufacture. If using entering air heat control, try to mount sensors midway of the heating coils and fully into the air flow, such that you are measuring the heat contribution off the coil. Occasionally I find sensors mounted between coil banks, presumably due to the convenient means of support, and wonder what the installer was thinking. Each kiln will undoubtedly be unique in its optimum sensor location, but you must exercise some common sense when positioning them. Keep in mind the prospect of heat radiation off the coils you are sensing. This radiation often results in erroneous readings and excessive valve cycling. If you cannot reposition the sensor, then shielding will often correct the problem, unless the sensor is so close to the heat source that reradiation from the shield becomes a problem. In this case, lining the shield with the ceramic fabric used to line boiler casings should take care of it.

For exiting air heat control combined with multizoning, it is imperative that temperature sensors be mounted as close to the load as possible for best results. Again, try to position sensors midway of the coil banks to accurately measure their heat contribution. Sensors mounted in the outside plenums must be mounted on swing arms that allow you to position sensors within 6 to 12 inches of the load. The sensors between the tracks in double track kilns need not be so close to the load, since airflow is largely horizontal in this area. Here, heat radiation off the center coils is the more important consideration, so sensors adjacent to the coils must be shielded from this heat. Note the discussion on shielding in the previous paragraph. Vertically, try to position the sensors as close to the middle of the coil bank as possible. This should also be approximately midpoint of your package as well. If the sensors align with crossouts or openings in the lumber, you will have to move them up or down to avoid the gaps. It is preferable to move them down, below the gap, to avoid the air from these openings.

When positioning sensors during the final preparations for running a kiln charge, pay close attention to the final position of the moveable sensors. They must not align with the gap between the ends of carts. Likewise, beware of air that slips under the load due to poorly fitting floor baffles. Control air that bypasses the load results in false high readings, which negate the benefits of multizoning and exiting air heat control.

Wet bulbs are mounted approximately 6 to 7 feet off the floor so that you do not bump your head when you walk down the plenums. They should extend 18 to 24 inches into the plenum. The bulb itself should be at least 1 inch off the water surface.

Entering Air Heat Control

Until 1982 all kilns were controlled using entering air heat control and the majority still are. These kilns measure and control set point temperatures of the air as it leaves the heating coils, but before it enters the lumber. All of the published drying schedules are set up for entering air heat control and are a throwback to the limitations of the original Filled System Kiln Controller-Recorders. These instruments indicate and record the high temperature side when configured with dual bulbs on each pen. On older kilns there were single bulbs on each heat zone pen. These single bulbs were commonly mounted in an airstream that bypassed the heat source. It is unclear what these systems were measuring, but they did use the same schedules. Indeed, it is likely that many of the original drying schedules were derived using this configuration. Using entering air heat control on a multizoned kiln completely eliminates any benefit to be derived from multizoning. You may as well have a single-zoned kiln.

Exiting Air Heat Control

Exiting air heat control was designed into the very first computerized multizone dry kiln. Although this was the first application for dry kilns, this method is widely used in related industries for drier control. Exiting air heat control senses and controls temperatures after the heated air has cooled in passing through the lumber, as it exits the load. This method automatically adjusts entering air temperatures to the heat demand of the lumber it passes through. High moisture content areas automatically receive more heat and dry faster than areas with lower moisture content. Exiting air heat control also automatically compensates for variations in air velocity, provided the kiln is properly multizoned. I am firmly convinced that we would not have enjoyed any of the benefits of the original multizoned kilns had we chosen to use entering air heat control over exiting air.

Why this works so well is related to the following formula:

Heat Delivery = f (Temperature **x** Air Velocity)

Distilled to its simplest form, lumber drying is delivering sufficient heat to our lumber to evaporate the water it holds. Ideally, if we could deliver heat evenly to our lumber, we would achieve even drying. This formula states that heat delivery is a function of the temperature and velocity of the air we pass through the lumber. We know that the air velocity through our lumber varies from location to location, say, top to bottom, within the kiln. If we use entering air heat control, we effectively fix the entering temperature to a single value. As such, areas of high air velocity will receive more heat and dry faster than areas of low velocity.

By using exiting heat control in combination with multizoning, the heat delivery in each zone automatically adjusts to variations in air velocity as well as the level of moisture in the lumber. In areas of high air velocity or low moisture content, exit air temperatures are maintained with lower entering air temperatures. Low air velocity and high moisture content areas require higher entering air temperatures to compensate. This ability to compensate for variations in heat demand within the kiln is the primary benefit of exiting air control.

Exiting air heat control requires that the published schedule set points be adjusted down to compensate for the naturally higher entering air temperatures that would result from using the schedules unaltered. How much they must be adjusted is a matter of experience. I will typically drop the published dry bulb temperatures 10°F or so on the first trial charge, all the while watching the venting during the early stages of drying. If the vents stay closed for a long period during the early stages of drying, I can likely increase the exiting air temperature. The published schedules for some species and items are so conservative that many of them may be used unaltered, although I do urge caution when doing this for the first time. It is far better to ease the schedule temperatures up slowly on subsequent charges, based on the results of current schedules, than run the risk of going too far and ruining a charge of valuable lumber. Make no mistake, you will most likely find yourself in new areas. depending on what you are attempting to dry with exiting air heat control.

One requirement of exiting air control is not as widely recognized as it should be. Since we measure and control temperatures as the air leaves the load, it is important to have uniform baffling and loading from one charge to the next. If the control sensors are located in the air that passes uncooled through crossouts, between adjacent units, under the load, or around the ends, then the control air is not representative of the lumber heat demand in that zone. Likewise, mills that sort two or more lengths into the same unit will have problems if the staggered ends align with sensors. These ends have less lumber volume and also demand less heat, which also results in a false high temperature reading. This concept is very important to the satisfactory operation of these kilns. Indeed, exiting air heat control is totally unsatisfactory unless this condition can be judiciously met. You may as well go back to entering air heat control.

Average Air Heat Control

Average air heat control is simply the average between the entering and exiting air temperatures in the kiln. This control method delivers some of the benefits of exiting air heat control that are not shared with entering air heat control. It has proven beneficial to kilns that have long air passes through the lumber load, i.e., package kilns. Using exiting air heat control on package kilns often results in extreme shifts in heat demand between fan reversals, which take a long time to settle out and likely impede drying progress. Rather than go back to entering air control you might give this option a try.

REFERENCES

1. Culpepper, L. G. "Application of Computerized Controls to Improve the Lumber Drying Operation." Seattle, WA: IUFRO Meeting, 1989.

Chapter Twelve

The Drying Process

Life shrinks or expands in
proportion to one's courage.
Anais Nin

Wood Structure and Its Relation to the Drying Process

If it is your goal to become an outstanding kiln operator, you must understand the structure of wood, how moisture is held in it, and how it is affected by the drying process. You do not need to know this information in order to run a kiln, but there is no way to excel in your job without this knowledge.

A Little Wood Technology

First of all, wood is dead. It was dead long before the logger delivered it to your yard and long before it was harvested in the forest. Take a walk out into the log yard and look closely at the cut end of a log. The most distinctive characteristics are the outer bark and the wood itself. Both of these are composed of dead cells. If you look closely at the area between the wood and the bark, you may distinguish the **cambium layer**. This is the living element of the tree trunk. The cambium layer produces cells of wood to the inside and cells of bark to the outside as the tree grows.

To the naked eye, the wood itself appears as light and dark concentric rings. Early in the growing season a tree is in a period of rapid growth owing to the availability of moisture and nutrients. The cells that are laid down during this period have very thin walls and appear as the lighter bands. This is known as **spring wood** or **early wood**. As we move into summer, competition for moisture and nutrients intensifies. The cells that are laid down during this period are darker in appearance, and their cell walls are much thicker. The dark portion of the ring is known as **summer wood** or **late wood**. Taken together these two bands constitute one **annual ring**. My dad, and I trust yours as well, showed me how to determine the age of a tree by counting the rings.

At the center of the concentric annual rings is the **pith**. This is the oldest portion of the stem and was laid down first as the tip of the growing tree elongated and reached toward the sky. The first seven to ten rings or so next to the pith is called **juvenile wood,** and these rings have slightly different properties, most notably in shrinkage, in comparison to the rest of the stem. Many folks attribute excessive twisting in their lumber to the presence of juvenile wood. It may be particularly noticeable when logging and sawing large volumes of plantation pine.

Softwood is a material composed of countless microscopic cells called **tracheids**. Tracheids are best described as closed-end hollow tubes whose lengths are approximately 100 times their diameter. Countless millions are arranged lengthwise in a tree stem and cemented together with lignin to form wood (figure 12.1). Think of them as straws all bundled together and oriented in the direction of the standing tree. A few cells are oriented differently. Locate the pith in a log and draw a line in any direction. The cells aligned in this direction are arranged into structures known as **rays**. Although the rays constitute only a small proportion (between 5 and 12% by volume depending on species) of the total number of cells, they are very important to softwood drying (figure 12.2).

Hardwoods are more complicated in their structure, although their components have functions similar to those of softwoods. The longitudinal

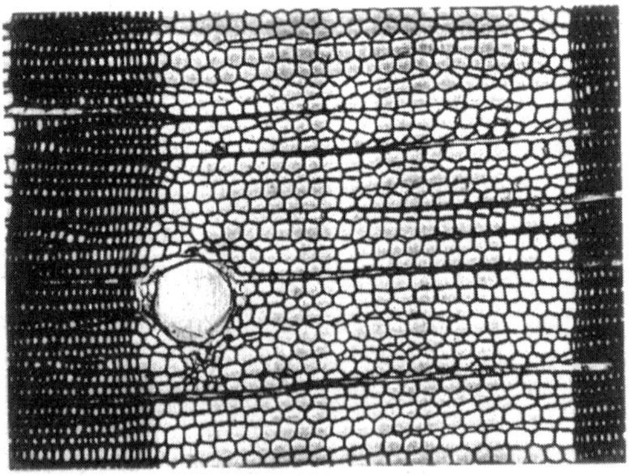

Figure 12.1 Microscopic cross section of softwood showing the tracheids. The thin-walled cells are known as early or spring wood. The darker, thicker-walled cells are known as late or summer wood. The large hole is a resin canal. Photo courtesy of Peter Koch.

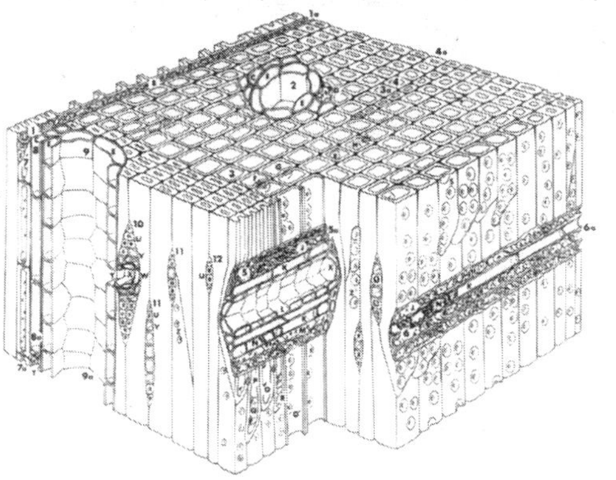

Figure 12.2 Schematic drawing of softwood. The majority of cells (longitudinal tracheids) are oriented vertically in the direction of tree growth. Rays are apparent on the exposed sides. Photo courtesy of Peter Koch.

cells are comprised of **vessels** and **fiber tracheids**. Although the rays generally make up a higher percentage of hardwood cells (between 5 and 32% by volume depending on species), they have fewer porous elements, being composed entirely of **parenchyma** cells. Indeed, in several species, such as white oak and black locust, the ray elements are plugged and severely impede the flow of water in this direction.

Between adjacent cells, water may move through openings in the cell wall known as **pits**. This is a very important avenue for moisture movement, although the openings may become clogged or otherwise closed as the wood matures or is dried.

Properties of Wood

Wood is **anisotropic**. This means that it has different properties depending on the direction in which it is tested. By contrast, materials such as steel, plastic, and aluminum are **isotropic**. There is no direction in these materials, their properties being the same in all directions. For wood, the movement of moisture and how it reacts by shrinking is affected by the orientation of the fibers in it. This is what we try to control as we dry our lumber.

There are three distinct directions in wood. The direction along the stem length is known as the **longitudinal**. Most of the tracheids are oriented in this direction. If you think about those straws, you can see that moisture would move fastest along their length. Now, imagine a short 1-inch section of straw. If you try to crush it between your fingers across its width, you will find that this is easily done. Crushing it along its length takes more effort.

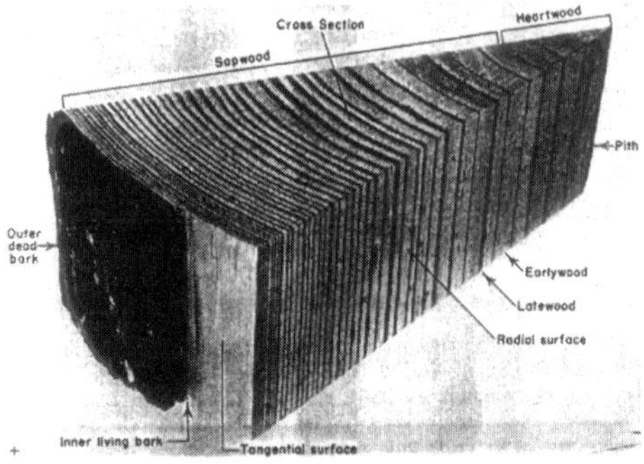

Figure 12.3 Wedge section cut from a softwood tree showing the primary parts of the stem. Photo courtesy of Peter Koch.

This is analogous to its shrinkage properties. Longitudinal moisture movement is fastest (12 to 15 times as fast as across the grain) and shrinkage the least (0.1 to 0.2%), all due to the orientation of the tracheids.

The second direction in wood is termed **radial** and is found by drawing a line drawn from the pith to the outside of a log. The presence of the rays in this direction means that moisture movement is intermediate in speed. Shrinkage averages 3 to 4%. Draw a line perpendicular to the radial direction. This is the **tangential** direction in wood. There are no cell structures oriented in this direction. For this reason, moisture movement is slowest of all the directions and shrinkage is greatest, approximately 2 times the shrinkage of the radial direction.

Go to the sawmill and pick up a green trim block in which the pith shows in the middle of one wide face. Compare it to a similar piece that has been through the kiln but not planed. You will notice that the dried board is thicker at the pith as compared to its edges, and yet it is obvious that it did not start out that way at the sawmill. Notice as you move across the board from the pith to the edge how the orientation of the rays changes. In the middle they are oriented through the thickness. At the edge they orient more to the width. It is this change in grain orientation that results in the variations in shrinkage. There is nothing you can do about it. If you place the green trim block on a shelf in the kiln control room and check it again in a month or so, it will look pretty much the same as the one that went through the kiln (figure 12.4).

While wandering in the log yard, you may have noticed that the center portion surrounding the pith of some logs was darker and surrounded by a lighter colored band. This darker portion is known as **heartwood** (figure 12.3). Its color comes from the presence of resinous extractives, gums,

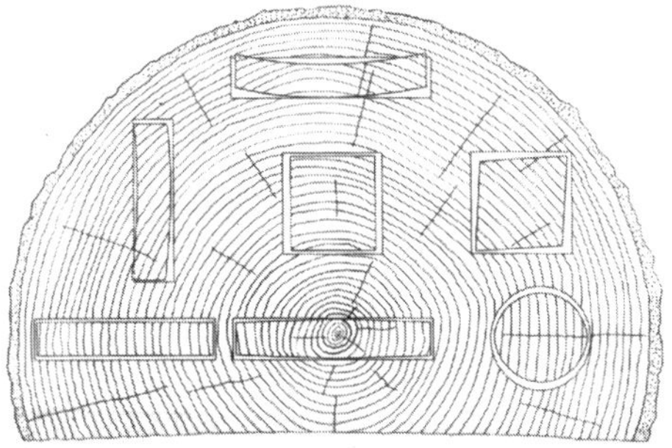

Figure 12.4 Illustration of how lumber changes its shape as it dries. The openings cut in the log end illustrate each board's green shape and compares it with its natural shape when dried. Note how the presence of rays in the radial direction influences shrinkage as compared to the tangential. Sketch courtesy of the U.S.D.A. Forest Service.

resins, and oils, which are deposited as the tree grows and ages. In the standing softwood tree, the moisture content of the heartwood might range from 30 to 50% MC, with a few species ranging to 100% MC. The heart of hardwoods has a relatively higher moisture content, ranging from around 50% MC up to 160% MC. In both soft- and hardwoods, the extractives plus microscopic structural changes in the heartwood render it less permeable to moisture movement. It dries slowly.

The lighter band of wood is known as **sapwood**. Its moisture content is very high in softwoods, ranging from 100% up to 250% MC, depending on the species. Hardwoods range between 50% and 150% MC, and it should be noted that there is generally less difference in moisture content between heartwood and sapwood of hardwoods than there is in softwoods. Lacking extractives, sapwood is very permeable to moisture movement and loses its water relatively quickly.

If you take a hypothetical softwood board composed of typically low moisture content heartwood and high moisture content sapwood and put them through the kiln, it would be very difficult to predict which one would reach an acceptable moisture content first, due to their differences in moisture permeability. It can go either way.

Another characteristic of wood further complicates the picture. This is the presence of **reaction wood**. It forms in leaning trees and aids the tree in righting itself. In softwoods, it is better known as **compression wood,** since it forms on the underside of limbs and leaning trees. In hardwoods, the term **tension wood** is used, since it forms on the top side. If you look around the log yard, you may notice a few elliptical logs where the pith is

Figure 12.5 The eccentric shape of this log indicates the presence of reaction wood. Since this is a pine log, we use the term compression wood to describe the area of wider rings on one side.

Figure 12.6 Edge board distortion due to the presence of compression wood. Compression wood is readily identifiable owing to the generally broad rings and darker color.

off center of the stem (figure 12.5). The wider side is the one with the reaction wood. The rings in this area are wider and darker than the surrounding normal wood. What makes reaction wood such a problem is that longitudinal shrinkage is four to eight times that of normal wood. If reaction wood appears along one edge of an otherwise normal board, the differences in shrinkage between the normal and reaction wood edges will cause the board to crook (figure 12.6). This is a sawing problem. It is desirable to orient the log through the system to either box the compression wood in the center of the board or confine it completely to one board, as is done in the new curve sawing systems. Once reaction wood gets to the dry kiln, there is nothing you can do about it.

Moisture Movement in Wood

Water is held in wood in two ways. **Free water** is the water found within the cell cavities **(lumen)** and comprises all of the water in wood above around 30% MC. The loss of free water does not result in shrinkage.

Bound water is bound within the cell wall and comprises the water in wood below approximately 30% MC. The removal of bound water results in shrinkage.

This approximate 30% MC threshold is called the **fiber saturation point (FSP),** and its actual moisture content varies with species. At FSP, all of the wood fibers are fully saturated with water, but no water is present within the cell cavities. This is an important marker for drying lumber. You might observe that in most published schedules the drying conditions are much milder above 30% MC than they are below. There are very good reasons for this, as we will see.

Three Mechanisms for the Removal of Water from Wood

Thermal expansion occurs in some kilns as they heat up rapidly to the initial drying set points and is typically seen in high-temperature kilns. I occasionally see it in high-performance conventional kilns as well. Here, free water is pushed forcibly out by the expansion of the water and air within the wood. Its action is not very important to overall drying, although I have seen upwards 1/2 inch of water deposited on the floor within the first three hours of a charge. If this water can be drained off, thermal expansion is a very cheap method of moisture removal but loses its momentum and importance as one nears drying temperatures.

Lately this water has become an increasing concern to those mills operating under zero discharge permits from their state environmental departments. These mills are forced to collect and dispose of all the water (including wet bulb discharge) coming from their kilns. It can no longer run out on the ground or to the nearest sewer. I have had reasonable success reducing, if not eliminating, this water by slowing the rate of heat rise. By doing this, water evaporates as it is pushed to the surface. However, this method does add an extra hour or so to the drying time. It is worth a try if you can spare the time. If this is an issue for your mill, and you are contemplating building new kilns, the floor may be designed to pool water so it can be evaporated from the floor rather than discharged to a drain. This entire issue promises to become worse rather than better.

The second mechanism for moisture removal is **capillary flow**. This is a rapid method of moisture removal and is highly dependent on air speed. It is the primary mechanism for the removal of free water and works through the cohesive and adhesive properties of water. **Cohesion** is the property where water molecules want to stick to each other. **Adhesion** is the property where water molecules tend to stick to other things. Both are considerations when setting up your schedules. In the early stages of drying it is desirable to set up a good capillary flow. This is done by raising the dry bulb temperature to a level that establishes and sustains the movement of water to the board surface. A wet bulb depression is maintained to remove this water from the surface. When properly done, as each water molecule evaporates from the board surface it pulls another molecule from the interior by cohesion. Be careful. Too wide a depression and we break up the capillary flow and dry out the surface. Wood is an excellent insulator, and this actually makes it harder for us to complete the drying process, since dry wood retards the movement of heat into the board. This increases the chances of surface checking as well. Narrowing the depression slows the drying process down from what it could be and allows water to pool on the board surface. This increases the probability of brown stain that is set up early in the cycle.

The adhesive property of wood comes into play only when drying a few species (redwood, white spruce, and others) that have relatively weak cell walls. In these species, it is possible to remove moisture so quickly from them that the cells literally **collapse**, destroying the shape, strength, and structure of the wood.

Below FSP, a third mechanism for moisture removal kicks in for the removal of bound water. **Diffusion** is the movement of water from areas of high concentration (the lumber core) to areas of low moisture content (the lumber surface). This migration is largely unaffected by airspeed and is the reason we are able to reduce fan speeds once we pass FSP. Recall, however, the discussion of this in the chapter on fan systems. There is usually too much variation in ending moisture contents to use this technique reliably on most softwood species, and it is not recommended. It is standard procedure for hardwoods.

The difference between the moisture content of the surface of a board and its core at the end of drying is known as a **moisture gradient**. There is always a distinctive difference between the moisture content of the surface and the interior at charge end. Next time you are checking lumber at the end of the drying cycle, drive the needles of a resistance-type meter slowly into the edge of a board as you watch the moisture display. If you are seeing more than 5 to 6% MC difference, then you may have a problem. We want to minimize the difference between surface and core to obtain the highest drying quality and can do so through the style of schedule we use. If you do not run equalization and conditioning steps at the end of the drying schedule, you must allow time for the lumber to cool and equalize within itself before sending it to the planer. Give it as much time as you possibly can.

Case-Hardening

Case-hardening is a drying defect generally of concern only to those drying wood that is to be remanufactured (ripped or resawed) either on site or by a customer. The internal stresses that develop in wood during drying cause it to cup when resawn and crook when ripped. This is generally only of concern to buyers of shop, moulding, furniture finish, and specialty grades. Occasionally it can be a problem for dimension grades and will be apparent on the landing table behind the planer. If you notice perfectly flat rough lumber on the planer feed table that cups or otherwise warps after it has been planed, it is likely case-hardened.

Imagine a board in the dry kiln. As we begin removing moisture from the wood surface we set up a moisture gradient within the board (figure 12.7). At some point the surface of the board begins to drop below fiber saturation point (FSP) and starts to shrink. However, the board surface is unable to shrink due to the higher moisture content wood below the surface. As the surface strains against the core, it begins to develop what is known as a **tension stress** loosely analogous to a stretched rubber band. If this surface tension stress exceeds the strength of the wood, then **surface checks** will begin to appear in the surface. The interior of the board is under **compressive stress**. Eventually the surface takes on what is known as a **tension set,** where the surface wood occupies more space than it would if it were allowed to shrink. Drying progresses.

Eventually the central core of the board begins to fall below FSP and starts to shrink but is prevented from doing so by the tension set in its surface. At this point the stresses begin to reverse. The inner core goes into tension that places the outer surface in compression as it combats the forces imposed on it. If the tension stress of the core exceeds the strength of the wood, **honeycomb** develops as small, lens-shaped checks in the board interior. Occasionally these checks are so severe that they extend clear to, and are apparent on, the surface. It is important to distinguish these interior checks from surface checks when tuning your schedules, since they occur at different times and for different reasons. Drying progresses.

The schedule ends and the lumber is removed from the kiln. In the end, the surface of the lumber is in compressive stress while the interior is under tension stress. The lumber may appear flat because the stresses are in balance with each other. Once resawn, ripped, or planed, we upset this balance, and the wood immediately warps as the stresses come into balance. The lumber is **case-hardened**.

If case-hardening is suspected, *and if it is important to your customers or you,* the standard tine test will detect it. To do this, obtain a representative board(s) from your load and crossout a wafer approximately 1 inch long, 1 foot from the end of the board. Using a band saw, cut along a line parallel to and approximately 1/4 inch below both wide surfaces from one end to approximately 3/4 inch or so of the opposite end. Knock out the center core.

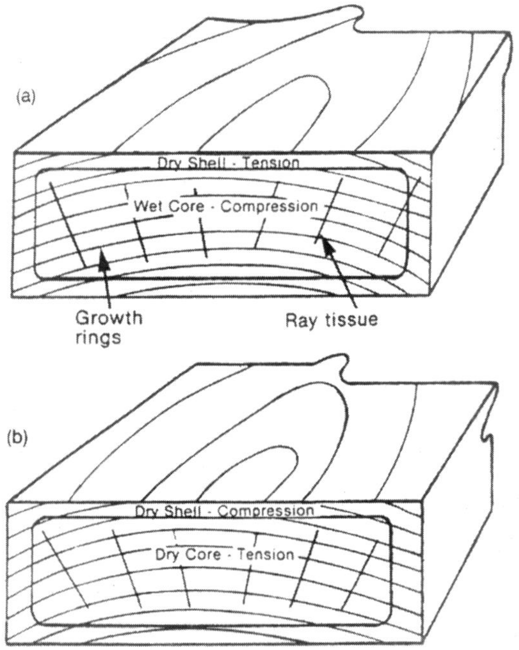

Figure 12.7 The development of drying stresses in the early stages of drying (a) and leading to casehardening at the end of the drying cycle (b). Sketch courtesy of the U.S.D.A. Forest Service.

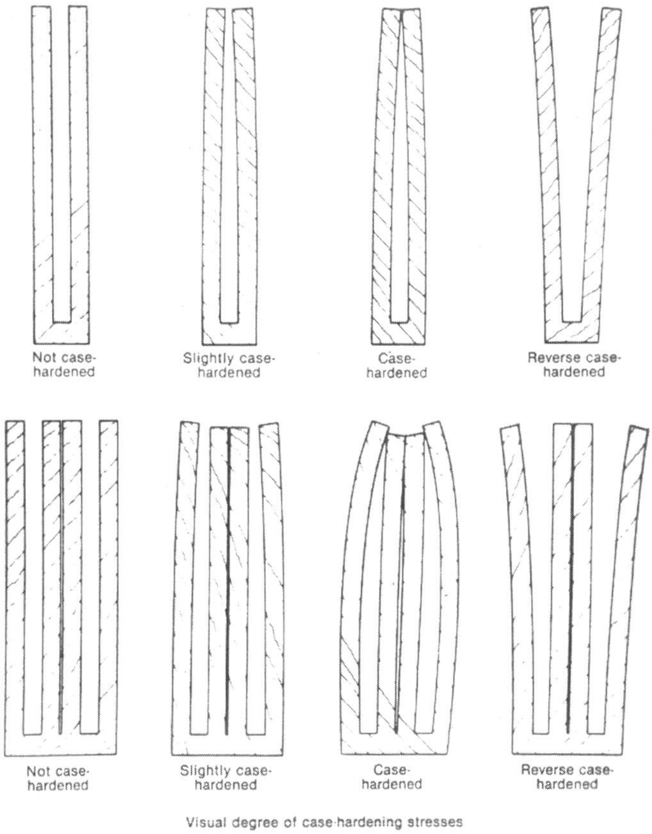

Figure 12.8 Standard tine test for evaluating the severity of casehardening. Sketch courtesy of the U.S.D.A. Forest Service.

How much the tines tend to pinch together is an indication of the severity of case-hardening in the sample. If the tines remain straight, pat yourself on the back. There is no case-hardening. Sample sufficient boards to get a feel for the entire load (figure 12.8).

Case-hardening is corrected by running a **conditioning** schedule as described in the section on humidity control. Follow the instructions, but be careful not to overcondition. This results in a situation known as **reverse case-hardening**. The tines bow outward when this has occurred. There is no remedy for this.

Developing Drying Schedules

This manual does not suggest specific drying schedules, since there are a multitude of published schedules available for every species and item you can list. Further, I am convinced that each kiln is unique with its own optimum schedule for each item you dry. As a starting point, I refer you to "Dry Kiln Schedules for Commercial Woods" published by the U.S.D.A. Forest Service.

This section is intended to aid you to develop the ideal schedules for your situation and needs, starting with the standard published schedules as a guideline. This is the final step after you have used the techniques of this book to optimize your heating, air, and humidity systems, since any changes here will impact on kiln performance and force you to start your schedule development from the beginning. Every kiln is an individual and must be treated as such.

Over the years I have used a variety of schedule styles and experimented with different philosophies in drying and at this point have settled on a few broad schedule outlines, which have given me good results over a broad range of softwood and hardwood species. Still, when I suggest a schedule for a new kiln or item, it remains an educated guess. I know that I will make multiple changes to the schedule depending on the analysis of the end results. How this is done is difficult to teach, but I will make the attempt here.

Embarking on a Kiln Improvement Program

Before we can begin, we must take the steps necessary to ensure that we are running our kilns at peak efficiency. We want our lumber manufacturing and handling practices down pat and the absolute best we can get them. Our kilns must be maintained and capable of performing at the limits of their ability. This is no time for fan outages, steam leaks, broken temperature sensors, or seized traps. Getting to this point has been the main focus of this book. If you have followed the recommendations and suggestions in each chapter, you will find that you are able to dry better lumber and dry it much faster than ever before.

I know, I know. Everyone has always told you that if you want quality you have to go slow and that if you want speed you must compromise on quality. I am here to tell you that this just isn't necessarily so. Sure, you can go out and crank up the temperatures in your kiln, dry a lot faster, and wreck a lot of lumber. But this misses the point. This book is about reducing the variability in your kiln process. If we do this, we can always push our kilns to their absolute limit without sacrificing quality. In fact, do not be surprised when quality actually improves.

The schedule you dry your lumber with is dependent on more than simply its species. Lumber sizes and the end users of your lumber heavily influence the selection of kiln schedules and whether or not you need to equalize and condition, as well. All of these are important, but you cannot even begin to select a schedule without considering kiln design, whether package or track, single or double; or drying process, whether conventional, accelerated, or high temperature. Keep in mind that what may be termed "high temperature" in northern areas does not begin to approach what is meant by "high temperature" in the South.

When first confronted with a kiln that is new to me I need to know species, size, and end user but will also pay close attention to the size of the heating and fan systems, length of air pass, the type of controls, and whether or not sprays are available for conditioning. If the kiln in question has been running for a while, I look at the schedules the mill has been using. Unless they are totally in left field, I start with them. Otherwise, I will rely on experience to dictate the structure of the schedule, based on all the above decision elements. If the species or item is new to my experience, I refer to the U.S.D.A. Drying Manual and compare its schedule to more familiar species. I know that I can always improve on any schedule in the book.

Keep in mind that all published schedules are based on entering air heat control. That is, they control the temperature of a heating coil based on the temperature of the air coming directly off the coil. My personal preference is to use exiting air heat control. My reasons for using this are outlined in the chapter on control systems. If your system allows you to use exiting air heat control, remember that you will have to reduce the dry bulb temperatures in the published schedules from 5°F to 20°F to compensate. On average air control systems the reduction would be less. Do not be surprised that as you tune your schedules and improve your drying operation you are able to dry with higher temperatures than you begin with. You may even discover that the published schedules are so conservative that you can use the standard entering air set points on exiting air systems with no difficulty. This will be particularly true if you are able to improve the efficiency of your fan system and deliver more air.

Now, at this point we need to run a schedule and dry some lumber. On new kilns, the powers that be must understand that there is always some risk at the beginning. If we have done our homework, the initial results should not be too bad. At least not bad enough to lose our jobs over. But first we must determine when the lumber is dry.

Determining When to Pull the Kiln

Kiln operators spend a great deal of gut-wrenching time trying to answer the all-important question of when to pull the kiln. They use a variety of methods. None are 100% reliable by themselves. The best results are achieved by using a combination of indicators.

When I first started kiln drying I met a gentleman who used his pocketknife to tell if the lumber was dry. First, he would tap the lumber with it, listening for the magic hollow sound of dry lumber. Next, he began cutting a sliver off the corner of a board and continued pulling it off with his fingers. How easily it pulled and how far were important indicators. Finally, he broke the sliver, listening for the snap and examining the broken ends. Fifty years of sawmilling experience told him if the lumber was "right" or not. The only problem with his method was that you had to have physical contact with the lumber. Modern kiln operators require more reliable methods for making shutdown decisions while the kiln is still running.

Over the years I have borrowed or developed many methods that I continue to use today. Since all kilns are different, it is no wonder that the signs you are looking for are different with each kiln. It is up to the operator to find the combination of methods that gives him the best results. Here is a listing of the various methods I have found useful:

1. Look at information available on your controls. Most computer systems today will give you information on temperature drop across the load (TDAL), wet bulb, vent position, total amount of energy going into the kiln, control valve opening, and the return and distribution duct temperatures on direct-fired kilns. From day to day you will gain a feel for the values that indicate when the lumber is dry.
2. Some kilns are equipped with indicators whose sole purpose is to determine the moisture content of the lumber. All indicators have their advocates. All have their limitations. They include resistance probes, electrical capacitance/impedance meters, acoustical emissions, and sonic meters, as well as load cell systems.
3. Take a look at the vapor escaping from around the doors or escaping around the vents. Note its color. Billowing clouds of steam may mean you need to run the kiln longer, although atmospheric conditions heavily influence this indicator.
4. Assuming the vapor is not billowing, carefully move your hand toward the vapor escaping around the door to test its dryness. If you feel adventurous and sure of yourself, carefully open an access door on the low-pressure side of the kiln and feel the airstream with your hand. The air swirling around dry lumber is not painfully hot.

WARNING: Placing your unprotected hand in the vapor flow around or within the doors and vents may cause scalding. This practice is not advised and must be approached with extreme caution and only then by individuals familiar with and willing to accept the personal risks. The same cautions apply to opening access doors while the kiln is running or entering a kiln soon after it is shut down. Keep low, and be prepared for a quick exit if the humidity is too high. While I may do these types of checks, no one has confirmed my sanity.

Once we have decided that the kiln charge seems dry, place it in hold or pause the drying. As your confidence grows, you may simply shut it down. Open the vents, access doors, and a main door on each end. The charge is likely wet if...

1. Water condenses on the steel above the main doors and rains down on you as you open them.
2. You notice a large number of small checks on the exposed ends of the lumber once you have opened the doors.
3. The pitch that has exuded to the surface and ends is still sticky.

Now, walk down the plenum areas on each side of the kilns.

1. Dry lumber has a distinctive odor. Try to recognize it for your kiln.
2. The kiln atmosphere has a distinct feel and appearance depending on how much moisture remains in the air. If you wear glasses, you may notice that they immediately fog. How fast they clear may be a good indicator for you.
3. The color of the dried lumber may be a good indicator, but keep in mind that color is heavily influenced by how long the lumber sat on the yard before it was put into the kiln and by the schedule severity. From charge to charge while running the same schedule, this indicator is reliable.
4. I often slip my hand into a few crossout spaces. If the board surface is too hot to touch, it may be wet.

Now, bring out the moisture meter. The one you use is purely a matter of personal preference. My only caution is that you be consistent from one charge to the next in how you use it. What you are trying to do here is to get a feel for how the moisture is running immediately after you have stopped the kiln. This is correlated to readings taken from the same lumber once it has cooled and is being processed through the planer. I discuss this technique in greater detail in the following chapter on quality control.

Every kiln is different, with its own personality, and it is for this reason that I am generally unable to determine when to pull kilns I am unfamiliar with. For this I rely on the operator (figure 12.9). If he says it is dry, then all that remains is the report back from the planing mill as to how good a job we did.

Figure 12.9 Big decision time. Do we pull it or not?

Analyzing the Results and Tuning the Schedule

Some drying defects are the result of poor handling and loading practices. Others are due to equipment deficiencies or failures. Still others are defects that result from some aspect of the drying schedule. It is important to distinguish them.

Handling and Loading Defects

These are easy to spot at the planing mill. Short, localized skips are due to poor sticker placement or quality (figure 12.10). Larger localized bows indicate that the placement of units on carts and crossouts must be improved. Do you find random wet spots down the length of your boards? This is likely due to stickers that are not laid straight across the units. You were supposed to fix these things before you started tuning your schedules.

Equipment Design Deficiencies and Malfunctions

Spotty wet lumber that always appears at the same spot in the loads is commonly due to poor baffle placement or design, sensor failure, trap or control valve failures, or fan outages. Wide variations in moisture content top to bottom or along the kiln length indicate variations in heat and/or air delivery. Entire spotty wet units mixed in with predominantly dry loads usually indicate differences in incoming moisture content. Wide moisture distributions are commonly caused by poor equipment maintenance and kiln design, not the schedule.

Figure 12.10 Short, skipped dress areas are a sure indication of poor stacking and handling practices.

If you are finding these problems, then it appears you need to start reading this book again from the beginning.

Drying Schedule Defects

Moisture Problems:

If you find a wide variation in moisture content across the width of the package, or you see a lot of warped boards on the edge or lops of your load, the ending dry bulb temperature is probably too high. Reducing side-to-side moisture variation is also accomplished by increasing airspeed, if possible.

If moisture distributions seem to be too wide, and the high moisture pieces are randomly distributed through the individual packages, try widening out the wet bulb depression a little earlier and certainly at the tail end of the charge. A lower beginning dry bulb temperature that gradually ramps to ending temperatures may also help correct the problem. If none of these appears to help, then an equalization schedule is always a possibility if the end use of the lumber justifies the time and expense. Remember, if the wet lumber always appears in specific areas of the kiln or stack, then the problem is one of kiln design, not schedule.

Other Defects:

Brown Stain Also known as coffee stain. This is a defect in finish, shop, and export grades. It is of no concern to dimension grades. It is most noticeable in the planed wood as a dark discoloration between the stickers (figure 12.11). The area under the stickers is usually bright and unaffected. You can also see it by shaving the rough lumber at the kiln with a pocketknife or small hand plane. Be careful using the hand plane, as its use will draw a customer's attention to what may be a minor brown stain problem, prompting a claim. I have seen a mill's brown stain problem disappear when it stopped using the hand plane. This can be a particular problem when shipping to certain European countries.

Brown stain is believed to be caused by the concentration of sugars and enzymes in the early drying stages, which becomes apparent as a dark discoloration as temperatures rise towards the end of the schedule. Control for the stain is done in the first 24 hours of drying and involves running as low a wet bulb depression as the material can stand without exceeding 140°F at any time during the schedule. The published schedules for susceptible species such as eastern white pine, ponderosa pine, and sugar pine address this problem and should be your starting point.

Southern yellow pine seems to react a bit differently. Here we look for a uniform golden color on the surface of the wood after it has dried. If the rough surface is very light or very dark in color, you may have a brown stain problem. It is necessary to drive the sugars to the surface in the early stages of drying to get that golden color. This confines the stain to the top 1/32 inch and is easily dressed off. Try running a little higher dry bulb set point early on to get the moisture flowing, adjusting the wet bulb set point

Figure 12.11 Brown or coffee stain often appears on the underside of the dressed board. Sticker placement is readily apparent. Severe brown stain may extend entirely through the board and cannot be planed out.

to prevent pooling water on the surface. In the past, while working on this species, I have lowered temperatures only to find that brown stain became deeper and more intense while surface checks deepened. You might also try switching from demand to conventional venting.

If the schedule you have been running successfully suddenly develops brown stain, look for the problem elsewhere. First check your equipment to see that everything is functioning as it should. Look at your log storage methods. Water-stored logs often have problems with brown stain and may be helped by aerating the pond. Try getting your logs into the mill and the lumber into the kilns a little faster. Only as a last resort should you begin fiddling with your schedule. After all, it worked well before. Something else has changed.

Checks, End and Surface These appear as small splits on the surface and ends of boards and are formed in the early stages of drying. Checks are commonly controlled by narrowing the wet bulb depression in the early stages of drying. Another option is to raise the initial dry bulb temperature while maintaining the same depression. This drives moisture to the surface faster, keeping the surface moist and helping to establish capillary flow. If lumber develops checks sitting on the yard, do not spray it during the early stages of drying to close them up. This only makes the checks wider and deeper. The lumber needs to be placed into the kiln faster.

Checks on the surface of 4 × 4s and thicker in which there is a boxed heart occur due to the differences in shrinkage rates between the radial and tangential directions. Some checking is inevitable, but it can be minimized by keeping ending humidity levels relatively high to reduce the moisture gradient. To minimize their width and depth, avoid overdrying of these items and allow plenty of time for the wood to equalize in the cooling sheds before dressing.

Splits, Roller These are splits that occur as cupped lumber is planed. They often extend the entire length of the board. I find that raising the dry bulb while keeping the wet bulb depression the same during the early portions of the schedule helps minimize roller splits by keeping the lumber surface moist and pliable. Many roller splits are not caused by cupping. If you dry common boards, those with heart pith frequently split end to end. This is caused by the differential shrinkage between the wood around the pith, which is oriented radially through its thickness, and that at its edges, which is oriented tangentially. The dried rough board is thicker in the middle compared to its edges and tends to flatten out as it passes through the planer. There is very little you can do for this beyond boxing this central pith into a thicker piece of lumber. Feeding boards through the planer pith side up may also help but will slow the planer down. Keep in mind that flat sawn lum-

Figure 12.12 Internal honeycomb caused by raising the dry bulb temperature too quickly towards the end of the schedule.

ber also has a tendency to cup due to the natural grain direction differences, although it is often possible to minimize its severity by following the above instructions.

Honeycomb This is the presence of small, lens-shaped internal checks, which are commonly aligned with the rays and form when the internal stresses within the board exceed its strength. Honeycomb is caused by raising the dry bulb temperature to too high a level before the core of the wood has fallen below FSP (figure 12.12). For softwood lumber, you should be passing FSP approximately two-thirds into the schedule. To control honeycomb, try reducing the dry bulb set point at that point in the schedule and ramping gradually up to it.

Warp The warp we are concerned with is...

- Crook—any deviation in straightness along the board edge. Possible causes: reaction wood, juvenile wood, localized grain distortion, and case-hardening.
- Bow—any deviation in flatness along the board length. Possible causes: reaction wood, juvenile wood, localized grain distortion, case-hardening, improper sticker or crossout placement, improper placement of units on carts, and broken stickers, cart supports, or crossouts.
- Cup—any deviation in flatness across the board width. Possible causes: flat sawn lumber, case-hardening, pith center, and low moisture content.
- Twist—the turning of the four corners of a board such that they are no longer in the same plane. Possible causes: spiral grain, localized grain distortion, and juvenile wood.

Warp has two main causes. Warp may be due to the physical characteristics of the wood that result in uneven shrinkage. These include the differences in shrinkage between tangential, radial, and longitudinal grain directions, as well as between normal and reaction wood. These also include the local grain distortions caused by changes in grain direction around knots or general spiral grain. Warp may also be due to a drying defect. It is important to distinguish the difference before attempting correction.

Examine the wood and try to determine if the occurrence of warp is random. Isolated boards point more to warp as a result of natural defects, and these should be examined closely for the cause. Occasionally the mill will get into a tract of timber with a high percentage of compression wood or be processing a smaller log size that tends to exhibit more warp. If you are using a drying schedule that suddenly begins warping lumber where it had always done a good job before, then you might suspect natural defects. Just make certain you rule out an equipment malfunction as the cause.

If warp is generalized throughout the load, then you should suspect the drying schedule as the culprit. Too wide an initial wet bulb depression or too low a beginning dry bulb temperature may be contributing to warp. If the warp seems mostly in the tops of carts or on the sides, then your ending dry bulb temperatures are probably too high. If the warp only appears in the dressed lumber rather than the rough, suspect case-hardening. Your schedule may have disrupted the capillary flow we want to establish and maintain early in the schedule. If there is no time to run a conditioning schedule, I find that exiting air heat control combined with ramping dry bulb temperatures minimizes case-hardening severity.

Always consider that you may be overdrying your lumber. The lower the moisture content of your lumber, the more it will shrink and intensify the natural shrinkage variations. If grade marking with a KD-19 stamp, try to aim for an average moisture content between 15% and 16% and a standard deviation around 2.5. If drying to KD-15, your average might be 12% or so with a standard deviation of 2.0 or less. By the rules, C&Btr kiln dried boards must have 95% of their pieces at 12% or lower. Here you may have to average 10% or less with a standard deviation of 1.0 to stay in compliance. We discuss this in greater detail in the following chapter on quality control. Obviously, how high your average can safely be is dependent on how much variation you have in your drying system. Wide variations will require that your averages be lower. Reducing that variation is really what this book is about.

One Final Note

Always go slow when tuning your schedules. Make modest changes one step at a time. If you are making a number of changes to schedules on each run, it is very difficult to determine which change had the intended effect or even if the effect you wanted was not offset by a change you made somewhere else. The time spent tuning schedules is a critical period. You must follow each charge through to the planing mill to verify quality and adjust each run accordingly. I like to keep one schedule for each item as my "best to date" schedule, which is always my fallback schedule if I have a problem. Each item might also have a "test" schedule where I am trying new ideas, always pushing the envelope.

The quote that opens this chapter exemplifies what excites me most about being a kiln operator. With the changes in drying technology that have occurred over the past decade, a whole new world has opened. Techniques that were tried and failed in the past often work well now because the technology exists to make it happen. I offer advice and a little bit of guidance in how to run your kilns because I want you, as a kiln operator, to experiment with new ideas and try new approaches. Indeed, the best kiln operators are often the ones who know nothing about drying at the outset but are willing to learn. They do not know what they cannot do. Mistakes will be made, but that is part of the process. This book is designed simply to help you to...

> Learn from the mistakes of others. You can't live long enough to make them all yourself. —Anon.
>
> But, then again...
>
> No Guts. No Glory.

REFERENCES

1. Brown, H. P., et al. *Textbook of Wood Technology—Vols. 1 and 2*. New York: McGraw-Hill, 1949.
2. Southern Pine Inspection Bureau, *Grading Rules—1977 Edition*. Pensacola, FL: SPIB, 1977.
3. Simpson, W. T., *Dry Kiln Operators Manual*. U.S.D.A. Forest Service, 1991.
4. Boone, R. S., et al. *Dry Kiln Schedules for Commercial Woods—Temperate and Tropical*. U.S.D.A. Forest Service, 1988.
5. Koch, Peter. *Utilization of the Southern Pines—Vols. 1 & 2*. U.S.D.A. Forest Service, 1972.

Chapter Thirteen

Quality Control

Experience tells you what to do;
confidence allows you to do it.
Stan Smith

Our Goal

The long-term goal of implementing a quality control program is fast, efficient drying coupled with the least possible degrade. Our immediate goal is to maintain the moisture content of our lumber at a level consistent with the grading rules and the needs of our customers.

Besides controlling moisture content, we may have other short-term goals. If we are starting up a new drying system, our first concern is that the equipment functions properly. Although we can inspect each part of the kiln during erection, there is always a breaking-in period as we test the kiln in operation, discovering and eliminating problems. If startup was delayed (it usually is), then we may have a large volume of lumber waiting to be dried. There is always pressure to get this lumber out, and speed may become the determining factor in how we operate our kilns. Just make certain the lumber is dry before it leaves your care.

The same goes for mills that experience extended kiln downtime or sudden surges in sawmill production. Faced with mounting green inventory, quality concerns often take a backseat to production. If the mill is always in a crisis situation, then serious steps must be taken to increase available kiln capacity. While you may ignore quality in the short term, always remember that the long-term survival of your mill depends on it.

Gaining control over your drying operation requires a concerted effort on your part. It will not happen on its own. This takes time, and the amount you need will depend on the size of your operation and condition of your equipment. To maintain your sanity it pays to establish a realistic time frame to complete the job, although it's doubtful that the job can ever be completed. There is always something else we can do to improve grade, decrease drying time, or just make things run a bit smoother. That is a good attitude to take and tends to make things more interesting in a field some find less than exciting. Watching a kiln run is kind of like watching paint dry or grass grow.

The kiln operator cannot do the job alone. He needs the cooperation of everyone from the sawmill to the planing mill to management. I have frequently entered a sawmill to find it divided into several seemingly autonomous areas—each one having its own limited area of concern to the total exclusion of others. Producing high-quality lumber must be everyone's concern, since each step in the process builds on and impacts the next. An observant and cooperative approach by everyone makes the job easier for all. The benefits are shared. Stress is reduced. We spend a substantial portion of our adult lives working. What a shame if we do not enjoy it.

How do you begin your kiln improvement program? You already have. By reading this book you have begun to gather the tools required to do the job. Now apply them. Walk through your mill to identify the items that are causing you problems with drying. Look at the stacked lumber ready for the kiln. Is it uniform in dimension? Are there distortions from poor stacking and setups? What is causing them? Is the kiln operating at peak performance? How can it be improved? By simply paying close attention to the details it is often possible to raise grade recovery by several percentage points, often at little or no expense. All you have to do is change the way you do things. To achieve maximum quality, you must gain and maintain control of your system. As a good friend once said, "A skilled kiln operator takes control of his dry kilns rather than letting them control him."

As you work on problem areas up to and including the kilns, you need some way to monitor your progress. How else will you know if your efforts have had any impact? Simply tracking daily grade recovery and making frequent moisture checks at the planing mill helps. The more reliable data you can get from the planing mill concerning grade and moisture, the easier it is to maintain control over your drying operation. This is an ongoing effort. Once the kilns are balanced and doing their best, things will still go wrong. Valves stick, traps seize, fans go out, and wet bulbs lose water. The list goes on. The planer mill says the lumber is wet. Is all the lumber wet, or just a few units? Where are those units coming from? The answers you receive tune your response by allowing you to hunt solutions with a rifle rather than a shotgun. Kiln operators who arbitrarily add or subtract time from their schedules in reaction to complaints of wet or overdry lumber from the planing mill are clearly out of control. They are doomed to swing from one extreme to the other. How can they possibly relax?

The Problem

In describing the problems drying southern pine, Peter Koch writes that "while all of the southern pines are relatively easy to dry, natural variability within the tree complicates the process. For example, the green moisture content of sapwood is higher than that of heartwood, but sapwood dries faster than heartwood. Because of this, either one may be overdried before the other reaches desired dryness." This statement can be applied to many soft- and hardwood species but is only one of the many interrelated factors that affect final moisture content distributions. These factors begin in the woods and extend throughout the manufacturing, handling, and drying process.

Beginning moisture content of logs is related to

- species
- tree age
- wood density
- geographical location (highland versus swamp)
- percentage of heartwood versus sapwood
- height within tree

Manufacturing and handling methods that affect moisture variation and drying rates include

- fresh-felled, length of dry land or water-storing logs
- method of sawing (quarter sawn versus flat sawn)
- variations in thickness
- variations in width
- variations in surface quality
- length of time before stacking
- length of time held on stickers before drying
- sticker thickness, orientation, and placement

Kiln factors that affect moisture variation and drying rates include

- condition of the kiln structure and panels
- condition, design, and efficiency of baffles
- variations in heat distributions
- variations in air velocity
- humidity control method

- drying energy source
- control method (exiting air versus entering air versus average air)
- kiln schedule
- drying time/shutdown method
- reliability of utilities (steam, water, air, electricity, fuel)
- use of equalizing and conditioning schedules

Finally, moisture variation is affected by the test method:

- testing hot versus cold
- where samples are taken (end, lumber side, lumber face, before/after surfacing)
- elapsed time between drying and testing
- tally method (handheld versus in-line meter)

In running any kiln drying system we want to gain as much control as possible over the various factors that affect our ending moisture contents. To do this we must recognize where to focus our efforts. Gaining the information to make changes to the system and determine how well we are doing in our efforts is the initial focus of the quality control (QC) program. Keeping tabs on the long-term performance of the system to head off problems is its primary purpose for being.

We have no control over the natural variation in incoming moisture content within logs coming to the mill, and it is pointless to try to gain any. However, mill management does have the ability to control variations due to storage methods and how logs are fed into the mill if they choose to exercise it. Manufacturing defects affecting drying are the responsibility of the sawmill, and they must be constantly reminded of this. Up to this point the kiln operator has little control, although a properly designed and implemented QC program will help apply pressure and obtain the justification to make changes up to this point.

Once the lumber clears the stacker it becomes your responsibility. How it is stored, staged into the dry kiln, and eventually dried cannot be shifted to anyone else. An effectively designed and executed QC program is your finger on the pulse of your operation.

Drying lumber to a uniform moisture content in most kilns is not a practical expectation, although given time, we could get pretty close. Variations in incoming moisture content along with variations in air and heat delivery conspire to widen moisture distributions. It is your responsibility as a kiln operator to identify the source of those variations and diminish if you cannot eliminate them. A properly designed and implemented QC program will assist you.

Moisture Distributions

If we sample the moisture content within individual units, carts, tracks or entire kiln loads, and plot each moisture content with its frequency of occurrence, we will end up with what is known as a **moisture distribution**. This moisture distribution most often, but not always, takes on the shape of a "bell curve" not unlike the illustration in figure 13.1. The highest part of this graph is often very near the **average moisture content** for the sample. As you move left and right along the scale, the line gradually drops as the number of high and low moisture content pieces diminishes. Along with the average moisture content, we need to know the **standard deviation** of our sample. Together, these two numbers describe the shape of the bell curve.

Standard deviation is a number without dimension. Ninety-five percent of the boards in our sample will be found within a range of ±2 standard deviations of the average for the sample. Obviously, we want this number to be as low as possible, as table 13.1 illustrates.

TABLE 13.1 RELATIONSHIP BETWEEN STANDARD DEVIATION AND MOISTURE CONTENT RANGE

Standard Deviation	Moisture Content Range*
0.75	3.0%
1.00	4.0%
1.50	6.0%
2.00	8.0%
3.00	12.0%
4.00	16.0%
5.00	20.0%

* of 95% of our sample

A very small change in standard deviation has a dramatic impact on the moisture variability of our lumber. As you become more familiar with its meaning and impact, you will observe that as you lower the average moisture content, the standard deviation also reduces. In fact, it is possible to obtain a standard deviation approaching zero simply by running your kiln until the lumber is bone dry, but you should probably consider other means of employment if you do so. Higher average moisture contents tend toward higher standard deviations unless you change the way your kiln dries. The Holy Grail of lumber drying is to raise the average moisture content closer to the upper limit of what the grade will permit while reducing its standard deviation.

Over time you may notice that the basic shape of the moisture distribution of the lumber produced by any individual kiln does not vary very much from charge to charge (figure 13.2). Obviously, variations in incoming mois-

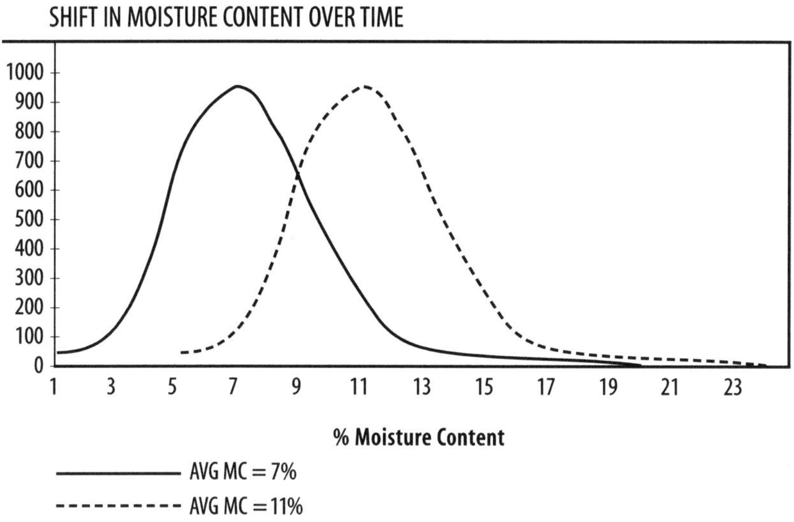

Figure 13.1 An example of a "bell curve" illustrating how moisture is distributed in a sample. Unless we mechanically alter the way the kiln dries, simply moving the average MC to the left by adding drying time only trades wet for overdry lumber.

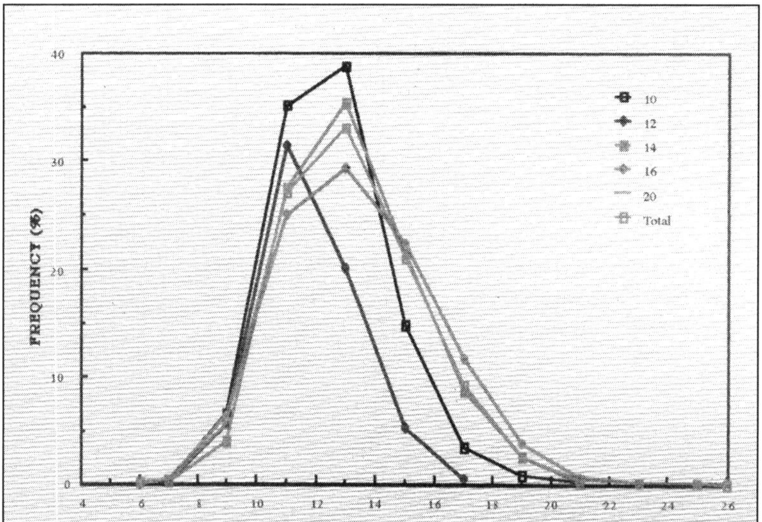

Figure 13.2 The moisture distributions from a charge of 2 × 6s. Note the similarity of the curves between the various lengths. Over time you will find that each kiln produces lumber that has its own distinctive moisture curve shape.

ture content or equipment failures will alter the shape, but excepting this, the day-to-day variation is generally very small. In the old days we dealt with the wet and overdry lumber we found at the planing mill simply by running the kiln for longer or shorter times in response to complaints. This simply shifts the average moisture content without altering the shape of the distribution. To make meaningful progress in our drying improvement program we have to alter the shape of the curve. Many kiln operators attempt to change the shape of their moisture distributions (narrow their standard deviations) solely through their schedules. While modest changes are possible, you must understand that the basic shape of your moisture distribution is determined by the design and efficiency of your drying equipment. Meaningful change in moisture distributions only comes by identifying and eliminating the source of the very high and very low moisture content pieces in our samples. Not coincidentally, there are specific places to look, and their elimination naturally reduces standard deviation.

Baseline Moisture Profile

When I am working at a mill I rarely have the luxury of conducting long, involved studies of the drying operation in order to determine what areas demand improvement. I start by walking through the log yards and sawmill while observing general practices and trends that impact on drying. I spend time in the green yard looking at handling practices. I go to the stacker after I have spent time in the planing mill observing a few packs of lumber as it is processed through the planer. I watch the personnel loading a kiln. I note the design and condition of the kiln and its various components and their geometry. Experience plays a huge role in what I do, but at some point management needs concrete proof of what needs to be done, if for no other reason than to justify to their superiors why they need to spend money on changes. This was the primary reason I started to use the **baseline moisture profile**. This technique quickly identifies the source of high and low moisture content boards and puts the information in a form that is easily understood by everyone (figure 13.3). This information may be gathered by hand on the yard (very laborious) or at the planing mill using a handheld moisture meter. Occasionally I find that the in-line moisture detector at the planing mill can be employed, which vastly speeds data collection.

From a kiln charge, select an entire cart of dried lumber that is uniform in color top to bottom. This gives you a reasonable assurance that the lumber in your sample entered the kiln as even-aged as possible. It is important to reduce as many variables as possible from your sample. Mark each unit as to location (top, middle, bottom) and let it cool in your normal manner. If checking moisture by hand out on the yard, record the moisture of each

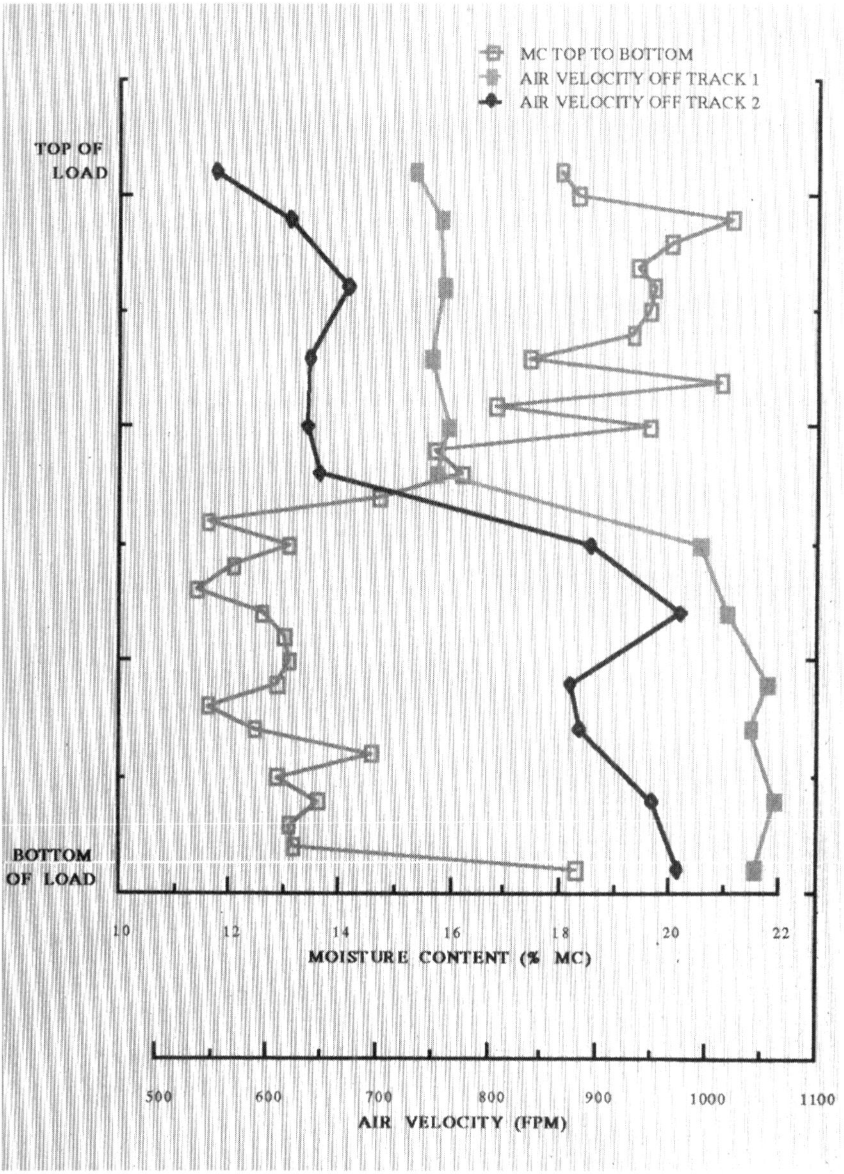

Figure 13.3 Comparison of moisture content top to bottom in a sample with the air velocity variation in the kiln. Note that high areas of air velocity typically result in areas of low moisture content. The air velocities between the two tracks are similar due to the presence of plenum baffles. These obviously need adjustment. Something is also wrong at the bottom of the stack.

board on each layer. If taking your sample at the planer breakdown, paint the ends of every other layer in each package on the end you will be testing (figure 13.4). This allows you to distinguish layers. As each painted layer slides off the breakdown, test and record as many pieces as possible from it. If you are collecting data with an in-line moisture detector, you may be able to distinguish layers by noting the drier edge boards, but this can often be difficult depending on species and the drying process employed. When finished, you should be able to reconstruct how the moisture content varies top to bottom within the cart. The primary advantage to going through these packages by hand on the yard is that you have a better sample and can reconstruct the moisture profile from side to side as well as top to bottom. Although you can collect a tremendous amount of data very quickly from an in-line moisture detector, it is often more difficult to glean useful information from it. Just keep in mind that what we are looking for is a snapshot of how the moisture is distributed in our kiln.

Calculate the average moisture content of each layer and draw a graph of how it varies top to bottom. At this point it is also useful to check the air velocity in the kiln to obtain information on how that varies. This information is graphed along with the moisture information. What you should find is that there is an inverse relationship between moisture content and air velocity. In other words, areas of high airflow will produce lumber that is lower in moisture content. Areas of low airflow will produce lumber that is higher in moisture content. This always applies to steam and thermal oil dry kilns. The correlation is less clear in direct-fired kilns and often does not apply. You may have to be content with studying the moisture content variation alone for clues to what is going on in direct-fired kilns.

Now, examine your data. In an ideal kiln there would be no variation in moisture content between layers top to bottom, and this is what we are striving for. In reality, you will find isolated layers that trend to high or low moisture content areas. This is where you must focus your attention. Pay particular attention to areas that do not follow the inverse relationship rule. Although this is a very unusual occurrence, it is not unheard of and points to severe imbalance problems within the kiln.

With the data in hand, go inside the loaded kiln and look around. The following problems have been found in the past, and I am always adding to the list.

Wet layers at the bottoms of carts may indicate…

- that floor baffles are contacting the lumber too far up, blocking layers.
- that the bottom runs of center heating coils are above the top of the kiln carts.
- there is no baffle below the center heating coils.
- there is a large gap between the overhead coils and outside wall.
- that the traps are undersized and waterlogging the bottom coils.
- poorly adjusted or designed plenum baffles.

Wet layers in the tops of carts may indicate...

- that the overhead baffles block the top layers. Overhead baffles should extend to the top edges of the loads only (figure 13.5).
- inadequate heat distribution top to bottom.
- poorly adjusted or designed plenum baffles.

Note that some kilns have naturally higher moisture contents in the top 2 to 5 layers.

Scattered wet layers may indicate...

- poor stick laying.
- short or broken stickers.
- poorly adjusted or designed plenum baffles.
- uneven heat distribution.

Overdry layers are often due to...

- poor heat control.
- poorly adjusted or designed plenum baffles.
- crossouts.

Wide moisture variation from edge to middle to edge indicates...

- air velocity is too low.
- dry bulb temperatures are too high.

Figure 13.4 If collecting profile data by hand at the planer, paint the ends of every other layer to distinguish them. Sample as many boards as possible from the painted layers.

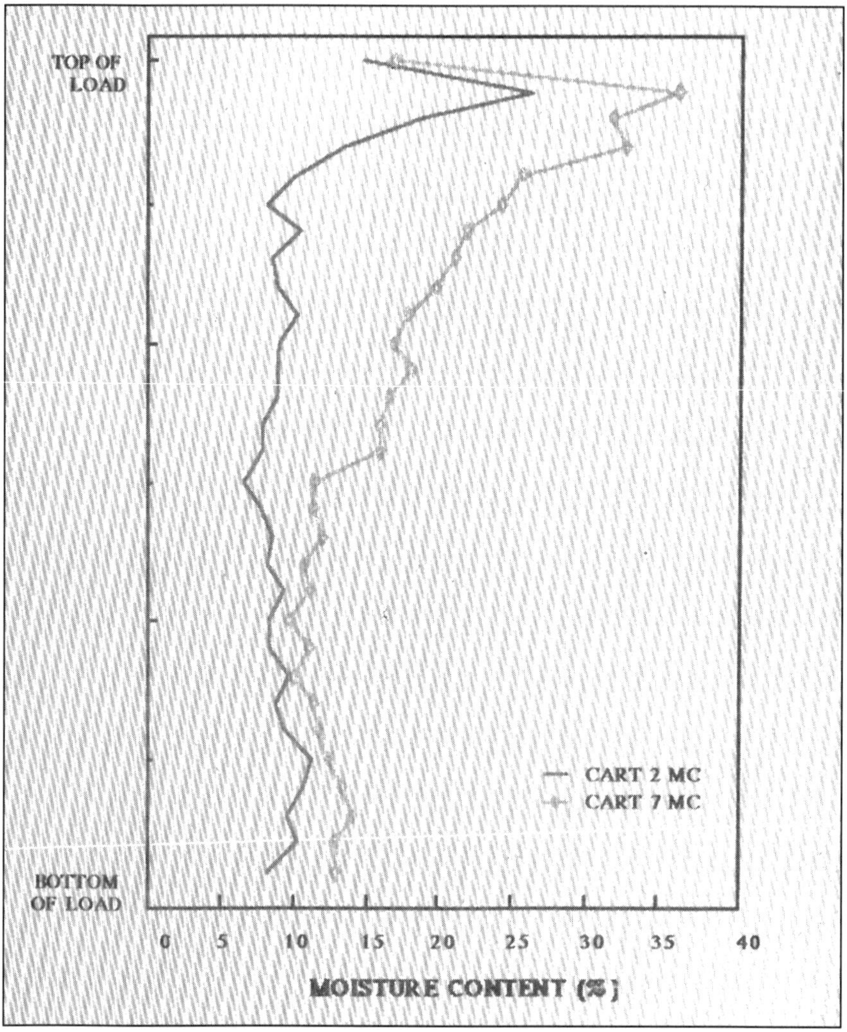

Figure 13.5 Variation in moisture content top to bottom in a direct-fired kiln. Data taken from second carts in from each end of the single track kiln. Note the variation along the kiln length due to the impact of the return air duct. The high moisture content areas at the top were due to overhead baffles that extended too far over the sides.

The point of this exercise is to identify the high and low moisture content areas of the kiln that are equipment related and eliminate them if at all possible. As will be seen, variations are often engineered into our dry kilns. Keep in mind that the engineers who designed your kiln may not have ever seen one in operation, much less studied them to understand what makes them tick. What looks like a good idea on paper does not always translate into good kiln design.

The ideal we are striving for is to reduce the variation in moisture content top to bottom to an absolute minimum. From side to side, the highest moisture content boards should be in the exact center of the pack. As you eliminate the high and lower moisture content boards from your sample, you will see your standard deviation become smaller. As standard deviations become smaller, you can safely raise your average moisture content closer to the maximum limit.

The baseline moisture profile is an extremely useful tool when adjusting plenum baffles for the first time. Adjust the baffles, run a charge, take a sample, and repeat the process until you have gotten the moisture content as even as possible from top to bottom. Once you are satisfied with it, remember to weld the baffles in place so you do not have to do it all over again when the maintenance man decides he knows exactly where they need to be.

Record Keeping

I am often reminded of a kiln operator I once worked with who had been drying lumber for a long time. Since I was new to this business, I often went to Willie to ask his advice about peculiar drying problems I was seeing at the planing mill or dry kilns. He always replied with one of two answers, which always began with him looking thoughtful and slowly rolling his eyes and rubbing his chin. "Well, you know," he would begin, "that always happens this time of the year." Or, "Well, you know, it really doesn't matter." It didn't take me long to start looking for my own answers, and I found that most of the problems I was seeing were not seasonal and that they often mattered a great deal. Something else was always going on. The only way to sort problems out is by keeping detailed records.

The more information you can keep on your kilns and the lumber they produce, the better. This gives you a solid base of information for future decisions. For example, I worked at a mill that ran two kilns with internal direct-drive fans. We were losing a lot of fan motors. Because we kept records of our losses, we were able to determine how much it was costing us and justify replacement with external fan drives. On a daily basis, complete information helps you track down the reasons for drying problems, as related to equipment, grade, moisture, and drying time.

Modern computerized controls that generate and store important process data for later retrieval are a tremendous help. I was once called into a mill to investigate the reason for extended drying times. We checked over all the systems in the kilns and could find only minor items, which did not account for the extra time. The computer system on the kilns tracked and stored the usual dry bulb, wet bulb, and moisture content data, but also tracked steam pressure. By plotting out the kiln charts over the previous month or so and laying them out on the floor, we were able to determine that something significant had happened at the boiler. This type of problem typically goes undetected with the kiln operator shouldering the blame.

Following is a listing of the type of information you may find useful at one point or another.

General information:

1. Service dates for preventive maintenance.
2. Information on emergency repairs, including dates, nature of the repair, and individual component (identify trap, fan, motor, bearing, etc.).
3. Interruptions of power, water, heat, or air.

Individual kiln charge information:

1. Items dried by size and grade.
2. Species.
3. Lumber location within the kiln (tag each lumber unit for later identification).
4. General quality of the green lumber (fresh, molded, etc.).
5. Schedule.
6. Drying time.
7. Turnaround time.
8. Final wet bulb reading at shutdown.
9. Final reading of any device used to determine when to pull the kiln.
10. General appearance.
11. Information from method used to determine moisture level at the kiln (hot meter readings, etc.). Average and standard deviation values are excellent.
12. Information from the planing mill on grade out and moisture content.
13. Kiln charts and any printouts.
14. Steam pressure and demand charts from the boiler.
15. Problems detected with the charge.

The more information you routinely gather from this list, the easier it is to run your kilns. You will be fully prepared when the question comes down as to why wet or overdry lumber inadvertently made it to the planing mill but more importantly, what was done about it. Try to design a standard form to fill in with the charge information and keep on file for each kiln for at least a year. Sooner or later you will need it. Remember, you are running the kilns. Don't let them run you.

Tracking Lumber through the Kiln

Item 3, under individual charge information, recommends tagging each unit of lumber so you can track it through the kilns to the planing mill. This is a very useful tool for isolating drying problems. If your mill uses inventory tags, then you are all set. Simply keep a sheet to record tag numbers and location within the kiln. Some mills print tags specifically for this purpose and include important information on them. Printed on heavy index card stock, they may brown, but I have never known one to catch fire inside the kiln. Use standard or plastic 1/4-inch staples to attach them at the stacker or before loading into the kiln (figure 13.6).

Several mills use tags perforated into three sections, with each section carrying the inventory number for that unit. As the lumber passes from the stacker to the planing mill, sections are detached to keep inventory up-to-date. At the stacker, the first section is detached with a record of lumber size and stacking date. At the kilns, the second tag is removed to track drying date and serves to locate the units within the kiln. Finally, the rest of the tag is removed at the planing mill. Information is added here on quality and moisture content. This allows you to track lumber throughout the system, accumulating much useful information along the way. This entire process is

Figure 13.6 Package tagged and ready to go into the kiln. Information includes thickness, width, length, stacking date, and day out of the kiln. Position in the kiln is entered by a predetermined code.

becoming increasingly automated, with tags designed to be read by bar coders tied into a computer system. These systems are primarily used to keep up-to-date inventories but also allow you to stage lumber into the kilns on a first-in/first-out basis. I highly recommend them.

Whatever system you use, try to spend a little time at the planing mill each day pulling tags and checking moisture content. Your visible concern about dried lumber quality will help foster the cooperation you need to keep the system running smoothly.

Collecting Moisture Information at the Softwood Kiln

The first chance you have to judge the quality of a charge occurs when you open the doors. Up to that point you have relied on the color or volume of vapor coming from the vents and doors to tell you that the lumber is dry. You may have monitored the wet bulb or have an in-kiln meter to judge dryness. Once inside, however, you can look at the color of the wood and breathe the vapors that tell the experienced eye and nose that the lumber is dry. Even so, it is only an educated opinion. It is time to bring out the moisture meter.

Resistance meters with 1-inch insulated needles are the ones most commonly found at dry kiln operations. There are also capacitance meters, which read moisture using a wide wand inserted between the stickers. Many mills swear by the latter, finding them fast and accurate. What is important here is that you use a meter in which you have supreme confidence consistently from charge to charge. If you try to use both meters, be prepared for discrepancies. They will also be at variance with in-line moisture detector data from the planing mill.

Meters with built-in temperature correction seem to only add to the confusion of using them. If your meter has this feature, set the temperature to 70°F and forget it. We will be using our meter simply to establish a correlation between moisture checks taken hot at the kiln and those at the planing mill after the lumber has cooled. Compensation for temperature is not necessary for this. I still recall the look of exasperation on the face of a kiln operator who had been told to subtract 25% from his hot meter readings to get the lumber moisture content. "How," he exclaimed, "do you subtract 25% from a meter that doesn't read?" That, my friend, is a very good question.

Store your meter in a box heated with a light bulb. This keeps the electronics warm and dry to maintain accuracy. If you keep it in an air-conditioned control room and take it outside or into the kiln, water condensation may render it worthless for your purpose.

Do not use short needles at the dry kiln. They are fine for the planing mill. Always drive the needles into the lumber so they align with the wood grain. Measuring across the grain alters the accuracy of the reading. Testing across the grain is acceptable only when probing the ends of units to determine the correlation between hot and cold readings. Drive the needles one-quarter to one-fifth of the board thickness using a smooth, relaxed, arcing motion. Although I typically do not use them, the protective blocks placed over the needles will help you to keep from breaking them. Go for full penetration only if you are interested in the moisture gradient.

Calibrate the meter before each use and change the batteries frequently. Always check calibration using standard calibration blocks available from the manufacturer. They are inexpensive insurance against equipment malfunction.

How you check your lumber at the kiln is a matter of personal preference (figure 13.7). Some mills check it within the kiln, first testing the exposed ends and then recording moisture from the sides of each unit.

Figure 13.7 Checking the moisture content of your lumber outside the kiln is one of many methods you can use to determine if it is dry, although it potentially entails more work should you have to push it back in for a little more time.

Other mills push the lumber out and test the sides as well as each end. Checking the ends has the advantage of allowing you the ability to check each board within a unit, not just those exposed on the sides, although most mills have difficulty safely gaining access to them.

As you probe each board, write its moisture content next to the test spot with a crayon. If you later decide to run the kiln a bit longer, you can easily compare readings to see how the moisture dropped. This is extremely useful information when trying to gauge how much longer you must run a future charge that is not quite dry.

The use of wands for checking moisture content is growing in popularity, and I find them particularly useful for seeing how the moisture varies from the edge to the side of the load. With these devices, make at least six checks top to bottom at each cart position, evenly spaced from as high as you can comfortably reach to the next to the bottom sticker opening. Although I normally take a single reading from the center of the pack, I have seen others take several readings across the width. Use the method you feel most comfortable with.

Checking the sides within the kiln is hard to beat in terms of speed. Keep in mind that the resistance meter information from the sides may not always reflect the true dryness of the lumber. If a fan goes out during the charge, the outside of a unit may check dry while the interior is wet. Still, it is a useful method as long as you understand its limitations.

Here are the factors you should consider when deciding on a sampling method:

1. Which boards are sampled and where? Do you check on the sides or ends? The package middle or edge? Look at the typical moisture patterns at the beginning of the chapter or break down a few units to see what you have.
2. When do you sample? In the kiln while it is still hot, or outside after a short cooling period?
3. How deep do the needles penetrate? Do we check full depth or use a block to prevent breakage?
4. How many boards do we sample? Do we exclude certain boards from our sample?
5. What results determine if we accept or reject individual units or even the entire charge?

Whichever method you use, stick with it. Try to establish a routine for moisture checking hot that all personnel follow from day to day. Remember, you are looking for an average moisture content reading for the hot lumber using *your* method, whatever it is. You may choose to ignore high moisture content boards if the boards on either side of them test dry. Never try to run your kilns so as to eliminate all high moisture content boards, or you will find the entire load too dry.

Your next task is to determine the correlation between these readings taken hot at the kiln and the final moisture content at the planing mill after the lumber has cooled. This allows you to refine your hot testing technique such that the hot test method becomes the final determinant of when the lumber is dry. If you attempt to run your kilns without immediate feedback but instead rely on planer mill data that is days or weeks old, you will be constantly chasing your tail and frequently out of control.

Remember that the hot check technique you use is important and must be standardized. Decide how you want to check your lumber and always check it in the same manner. Changing anything affects the reliability of the technique. If you have an established technique and change the design of your kilns, particularly the control method (exiting versus entering air), expect your testing results to change. Unfortunately, there are no guidelines to help you anticipate what the change will be.

Changing any component of the test procedure affects the results of the testing method. Be consistent in your testing and record your results. Over time you will tune your technique, adding or subtracting aspects to improve your confidence. For example, some mills check the outside of their units within the kiln. They drive the needles full depth, avoiding heavily pitched or wide-grained pieces. They have found that they are in good shape as long as they get no readings over 18%. One or two percent above this, and they will put a little more time on it. They know from experience how long that will be. Our mill, on the other hand, tested both unit ends outside the kiln after the lumber had cooled for 15 minutes. Heavy pitch was ignored. Penetration was limited to what was attained with one smooth motion. We looked for the majority of our readings to be around 15%. Choose your own method but stick to it.

Collecting Moisture Information at the Hardwood Kiln

Moisture is typically monitored far more closely in the hardwood kiln than it is in any softwood kiln I have ever visited. The reason is simple—the value added to their product by drying and the sensitivity of many of the hardwood species to drying defects demand it. There is far too much to lose. Also, most hardwood species are dried based on moisture, not time, although many of the soft hardwood species can be monitored using the methods already described in this book.

For hardwood mills with exacting standards, two methods are employed to monitor moisture content in the running kiln. Historically, moisture has been monitored through the use of sample boards, which are weighed daily to determine what the drying conditions should be. The procedure for setting up this system is well described in the U.S.D.A. *Kiln Operator's Manual,* and I will not go into it here. This has been the standard method for what seems like the beginning of time.

Over the last few decades the use of resistance probes, which are not unlike a resistance moisture meter, has gained increasing acceptance to the point of being commonplace. Their chief advantage is the availability of real-time moisture data without having to enter the kiln. Each kiln is equipped with six to eight or more moisture probes, which are inserted by the operator as the kiln is loaded. The information is used in the same manner to adjust drying set points, and many of the systems will automatically advance the kiln set points in response to the changes in moisture content.

Hardwood kilns are typically operated based on the slowest drying boards or areas of the kiln. I recommend that the probes be inserted into representative boards in units that are positioned in the slowest drying areas of the kiln. Most hardwood kilns are package kilns, and the slowest drying areas are typically the bottom center units in the middle line of lumber. For track kilns, the slowest drying areas are typically the bottommost units on a cart. For any operation, the use of the baseline moisture profile will help you identify the best area for sensor placement, as well as suggest areas for improvement in the kiln design.

Some manufacturers of these resistance systems recommend placing the sensors into the center of packs, forcing the operator to laboriously split the units to insert the probes. Although this may be the absolute best location for the probes, it severely increases your turnaround time. You will likely find only 1 to 2% MC difference between the readings you get from this method versus simply inserting them into the edge boards. This is an insignificant difference. Run a test on a safe species to confirm this for yourself.

At charge end, following equalization and conditioning treatments, the typical hardwood kiln operator has amassed an impressive amount of moisture information. A quick check with the handheld meter, and the lumber is often on its way to the breakdown for grading and shipment. Still, he needs to follow up on the lumber before it leaves the mill to make certain all is fine.

Monitoring Moisture Content at the Planing Mill

Unless you are obtaining moisture content information from the planing mill, you are still leaving too much to chance. For one thing, you must periodically affirm that the hot check readings remain reliable gauges of the ending moisture content. Many mills assign the grader to check moisture content. Although they are the logical people to do this, since they see all the lumber, the probability of bias in the data is great. Graders often check boards only when they suspect wet lumber or to confirm that the lumber is "burned up." This information, although it serves the graders' purposes, is worthless to the kiln operator unless you can determine the specific kiln charge it was dried in and the location of the lumber within that charge. It is not representative of the lumber as a whole. Dependence on this data collecting method is very risky, although a few mills do not have any alternative. Even if you collect moisture data yourself, you will not see all of the lumber. There is also a tendency toward bias in the other direction.

How do you eliminate bias errors from checking lumber by hand? You cannot, although you can reduce it by limiting the ability to arbitrarily select lumber for testing. One method is to preselect boards by spray painting a diagonal stripe on the inspection ends of your units. Start in the upper right corner and spray a line to the lower left. On the grading chain only the painted boards are tested and recorded. You might have the graders check 10 or more boards an hour. If this is too tedious, set up a timer with a light that comes on at preset intervals and hang it in the grader's view. When the light comes on he begins checking end-painted boards or boards at random until it goes out. A few minutes once an hour is plenty. By the end of the day, you will have data from several hundred boards, which will give you an overall picture of what is going on, but there is still no reliable way to correlate this information back to specific charges.

The most precise tool for maintaining moisture control over the dry kiln is the in-line moisture detector. These machines sample and tally each board automatically, eliminating the bias errors of human sampling. Most detectors allow you to sample individual units to compare lumber from different areas of the kiln or between kilns. I would not attempt to run a high-production operation without one. To obtain maximum benefit, you must obtain your data in a logical manner. Ideally, you want data collected and stored from individual packages, which would allow you to reconstruct the moisture profile for an entire kiln charge. Simply collecting moisture data from shift beginning to shift end does not give you any useful data, since it will represent multiple kiln charges. To control your kilns you must have specific information.

The most logical place to install an in-line moisture detector is ahead of the planer, such that high moisture content boards can be dropped out prior to dressing. This also allows you to more tightly control the data being fed to the computer and assures you that the data you get is indeed from specific packages. Most mills install these detectors on the outfeed of the planer, which is most convenient and reduces equipment costs, although here it is simply a data collection device. Collecting this information ahead of the planer gives you the data and allows you to recover the high moisture content boards by putting them back through the planer after they have been further dried.

Planer operations with computerized trimmers and sorters produce detailed information on grade recovery and trim. This information is difficult to compare from one charge to the next and is not suitable for this purpose. However, it will tell you how you are progressing with your kiln improvement program over long periods of time.

A useful tool is to have your agency lumber inspector go through a package of your outs every month and detail the reason why each board appeared in the grade. This information is then categorized into the various types of defects: natural, manufacturing, and drying. I further break the drying defects into those that are due to handling and those that are due to the equipment (table 13.2).

TABLE 13.2 GRADE-LOWERING DEFECTS

NATURAL	MANUFACTURING	DRYING HANDLING	DRYING DRY KILN
Compression wood	Torn grain	Warp	Checks
Decay	Stain	Kiln kink	Internal
Sloped grain	Variations in thickness	Broken boards	Surface
Knots	Variations in width	Sticker depressions	Checked knots
Holes	Rabbeted edge	Stain	High/low MC
Heavy pitch	Offset saws	Splits	Splits
Pockets	Splits	Skip dressing	Holes
Shake	Saw cuts		Brown stain
Wane	Wane		Warp
Cross breaks			Crook
Stain			Bow
			Twist
			Cup
			Collapse

Several defects appear in multiple categories, and it is important for the inspector to distinguish between them. Stain can develop in the standing tree due to tree injury, in the log yard due to improper storage procedures, in the green storage yard, or in the kiln. A skilled inspector will determine the source and properly allocate the defects. This information actually gives the mill a snapshot of how the operation as a whole is performing and helps it to focus on the areas that require the most attention. Graphed over time, one can track how the different areas of the mill are progressing or falling

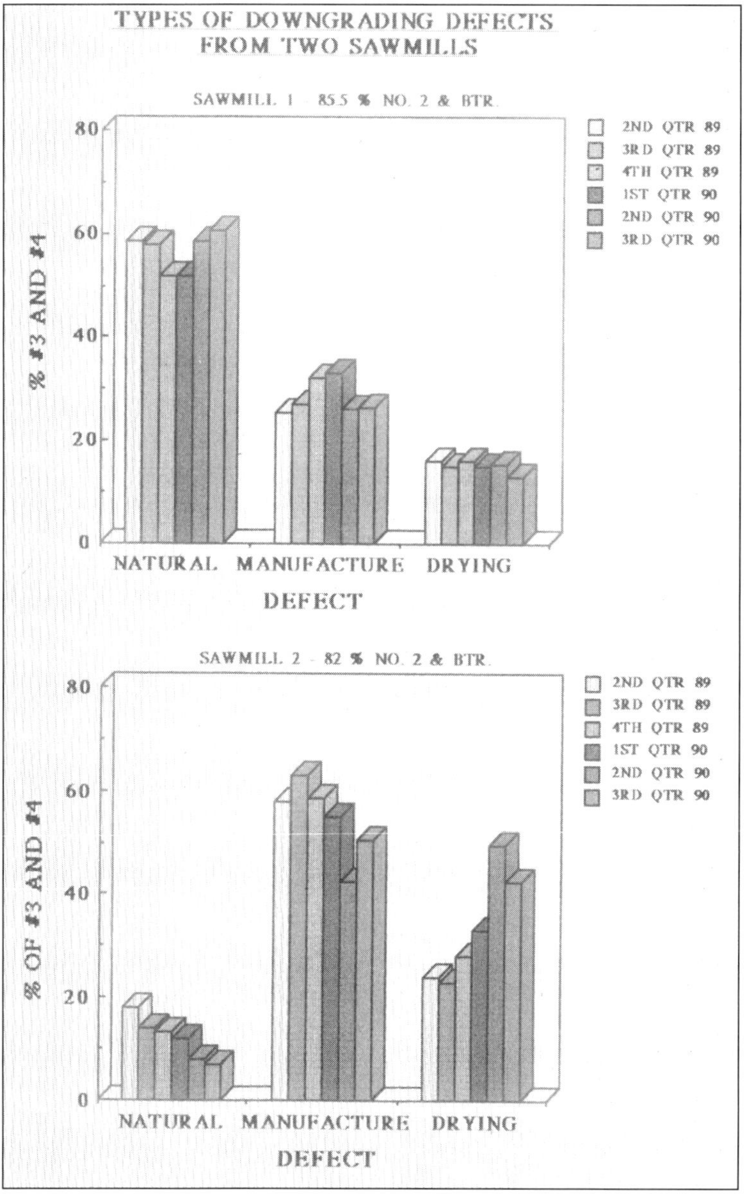

Figure 13.8 Keeping track of the various defects in your lower grades may alert you to shifts in quality, as well as give you some indication of where to look for remedies. This graph compares data from two separate mills.

behind. At an ideal sawmill, all of the boards would be in the grade due to natural defects (figure 13.8).

Applying Quality Information to the Kiln

The first rule is: Never overreact. When the planer supervisor starts screaming about wet or overdry lumber, look at your information. If you have kept good records, it will be easy for you to track down the solution. Was this an isolated unit, or was the entire charge wet? Was there an equipment malfunction?

Be very reluctant to make drastic changes in your schedules. After all, they worked fine before. Nine times out of ten you will discover a mechanical problem in the kiln that contributed to the problem. If you have done your job by identifying and eliminating the sources of wet and overdry lumber within your kilns and can verify that there were no malfunctions, only then should you consider altering the schedule. Always keep in mind that there are a multitude of problems that can contribute to the problem before you even placed the lumber into the kiln (figure 13.9). Tamper with your schedules only as a last resort.

Examine the information available to you. Get the planer personnel involved in determining the extent of the problem. Finally, make an informed decision.

Figure 13.9 Simple observation is a valuable tool to identify drying problems. Although saw offset is an obvious problem, here the darker marks indicate that stacking and kiln setups have room for improvement as well.

REFERENCES

1. Grant, E. L. *Statistical Quality Control.* New York: McGraw-Hill, 1952.
2. Koch, Peter. *Utilization of the Southern Pines. Vols. 1 and 2,* Agricultural Handbook No. 420, Forest Service, Supt. of Documents, GPO Stock No. 0100-1483, 1972.
3. Southern Pine Inspection Bureau. *Grading Rules—1977 Edition.* Pensacola, FL: SPIB, 1977.

Chapter Fourteen

Kiln Maintenance and Design

Discovery consists in seeing what everyone else has seen and thinking what no one else has thought.

Albert Szent-Gyorgi

Preventive Maintenance

An important part of running a successful drying operation is understanding the drying process well enough to recognize when something has gone wrong. Unless you are very observant of things that are out of place, the best way to accomplish this is by establishing a preventive maintenance program. Rather than react to sudden equipment failures, periodically check each system component and make routine repairs as required. It is much easier to schedule downtime than to catch up following unexpected emergency repairs. Worse, some problems are slow, progressive failures, which take some time to detect. Head them off by checking each part of the kiln on a regular basis. It should be emphasized that the better you are able to maintain your drying equipment, the lower your drying costs. Not just in the obvious areas such as energy usage but also in terms of the quality of your dried product.

The checklists at the end of the chapter are designed to walk you through the kilns, checking the important components that assure reliable operation. One checklist covers steam-heated kilns, the other direct-fired. Modify one of these to suit your specific needs.

You should set up several service logs to track the most critical components of your kilns. These might include a fan motor log to track the loss of motors. This will be useful to determine if certain motors are failing on a regular basis, whether they are new or rewound, and how much life you are getting from them. You might set your traps up on a preventive maintenance program where individual traps are routinely rebuilt every two to three years, depending on service severity. You will need to tag each trap to identify it, as well as keep the service dates organized in a log. In short, anything that requires regular maintenance should be logged, as well as any failures of any kind. Some mills simply keep a daily diary for this purpose, rather than maintain logs for individual components. The information is there when you need it, and it is easier to keep. It simply takes you a little longer to draw out the information you are looking for when the time comes. Whatever method you prefer, the bottom line is—*keep records*. It is hard to judge which direction you need to be heading in if you are not certain where you have been.

Analyzing Kiln Charts

Dry bulb

Symptom	***Possible Cause***
Dry bulb above set point	1. Control valve stuck open. 2. Control valve bleeding through due to a) worn control valve seats, or b) I/P transducers are out of calibration. 3. Heat entering from an attached kiln. 4. Bypass valve left open.
Dry bulb below set point	1. Control valve frozen closed. 2. I/P transducer failure. 3. Hole in control valve diaphragm or air line. 4. Vent(s) stuck open. 5. Loss of steam pressure. 6. Loss of air pressure. 7. Controller malfunction. 8. Trap failure or blocked drain lines. 9. Boiler pressure exceeds upper pressure limit for installed trap orifice.
Dry bulb shifts with fan reversals	1. One vent row is stuck open. 2. Dry bulb sensor positions are not symmetric from one side of the kiln to the other. 3. Poor baffles or gaps between the lumber units allow air to escape around lumber and hit heat sensor on exiting air systems.

4. Very different airflow from one fan direction to the other.
5. One sensor is out of calibration.
6. Only one sensor is being read due to a broken relay on an electronic recorder/controller.
7. One capillary is crushed on a dual-bulb vapor-bulb controller.
8. In multizone systems, one or more dry bulbs may have failed or be out of calibration.

Dry bulb oscillates about set point

1. Heat radiation on the temperature sensor is causing excessive cycling.
2. Controller PID algorithm needs tuning.
3. Steam pressure is oscillating.

Wet bulb

Symptom / ***Possible Cause***

Wet bulb above set point

1. Vents are not functioning, check
   a. diaphragm or positioner for leaks in air line.
   b. for a malfunctioning controller or I/P transducer.
2. Large steam leak.
3. Encrusted wet bulb sock, a sock that has dried out from lack of water, or a missing sock.
4. Steam sprays that are stuck open.
5. Inadequate vent capacity—you may need to add more vents.

Wet bulb below set point

1. Vents stuck or blown open.
2. Insufficient moisture in kiln to maintain set point.
3. Sprays not coming on due to blown valve or airline.
4. Wet bulb sensor is physically laying in the water.

Wet bulb shifts with fan reversals	1. One vent row is not coming open. 2. Moisture is seeping in from an adjacent kiln. 3. Steam leak. 4. One sock is encrusting on a dual wet bulb system. 5. On dual wet bulb sensor units one of the sensors may be damaged or out of calibration. 6. Ambient air may be being drawn into kiln environment in the area of the wet bulb in one fan direction and not the other.
Wet bulb oscillates around set point	1. Improper venting. 2. Vent controller PID needs tuning. 3. Vent positioners have too much play in them.

Fire Prevention

The following is excerpted from "Fire in the Dry Kiln" by Kent Townsend:[1]

Fire in a dry kiln can be a disastrous occurrence. Although there are some reported incidents of fires in steam kilns most fires occur in direct fired kilns and most often in those kilns using a wood fired suspension burner.

Fires that result from wood fueled suspension burners are not that infrequent, especially on poorly cleaned and maintained burners. Mills that have these types of kilns and burners soon realize that close observation of the kiln during operation is necessary in order to detect fires quickly and take prompt action. The importance of periodic cleaning of accumulated material in the burner, adjustment of the fuel/air ratio to manufacturers' specifications, and the repair of refractory and general burner equipment are all very critical to the prevention of fires in the dry kiln.

Fires seem to begin from two major causes: 1) uncombusted fuel which causes sparking, 2) leaks in the air distribution duct.

The most successful way to [fight a fire in the kiln] is to first be prepared and have a plan before the fire ever occurs. Discuss the plan with the kiln people and/or those who will fight the fire. When the fire is discovered, shut down the kiln including fans, burner, vents and doors, sound the alarm, call for assistance, and activate the sprinkler if not already activated. Once assistance arrives prepare the fire hoses and fire system to extinguish the fire once it is outside the kiln. Once everything and everyone is ready, push the load out towards the green end with a forklift. Extinguish the fire. Some common sense regarding the size of the fire should be exercised before attempting to push it out. If it is too large it could endanger the life and safety of the fork lift driver in which case the fire will have to be fought in the kiln. Most fires of this nature, however, give ample time for careful thought and action if the kiln itself is not burning.

Other sources of fire can be eliminated through common sense. Never allow dry lumber to remain in a kiln for longer than a few hours, particularly in those kilns drying highly resinous species such as pine. Always push it out as soon as possible at charge completion. You should also be on the lookout for problems if you are drying heavy timbers in a direct-fired wood waste kiln. The heat demands for thick timbers are often far below the turndown on the burner, and timbers end up running for long periods of time on low fire. Many of these burners spark excessively on low fire and may start a fire at the tail end of a schedule when the wood is driest.

If you are installing sprinklers in your kiln, they should be dry pipe rather than wet pipe systems. Wet pipe systems tend to deteriorate quickly in the kiln environment.

Purchasing a New Kiln? Some Design Tips for Easier Kiln Operation

Part of my business is designing kilns and drying systems that accomplish their intended purpose. Not only must they be capable of drying high-quality lumber, they must also be easy to use. With a little thought you can avoid many of the chronic problems that plague dry kiln operations. Take fan maintenance as an example. If maintenance personnel must squeeze through small access doors or clamber over or through pipes and fans in a hot environment, (1) they will be reluctant to do their jobs because of the physical discomfort; and (2) they will likely rush through their work and miss something. Give them easy access, cool working conditions, and plenty of light, and there is a better chance that the work will be done right. It may cost a little more up front, but the reward is a drying system that operates at higher efficiencies. This translates into dollars over the long haul.

If you are considering purchasing a new dry kiln for your operation, you might want to consider a few of the things I look for in a dry kiln. I also have specific preferences for how particular pieces of equipment are installed, which are scattered throughout the text. I will try to compile them here, but keep in mind some may simply be personal preferences. When evaluating any option or feature keep in mind that you want to end up with a kiln that is not only simple to operate but easy to maintain as well. Let's go through the items I have grown to prefer over the past 20 years of specifying, building, running, and maintaining dry kilns.

Foundations, Tracks, Carts, and Doors

Raise the perimeter foundations on which the columns rest a foot or so off the floor, particularly in kilns with steel structures. Since water often pools on the floor in the early stages of drying, there is a tendency for the bases of these columns to rust out if built on a flat slab. Supply drainage inside of the kiln to remove excess water.

Bury tracks in concrete up to the ball. I used to leave the ball of the rail standing about 3/4 inch above the supporting slab. A few years back Mr. Bill Shaw showed me another way. Prior to pouring the slab attach, 1/2 × 1-inch strips of foam insulation to both sides of the track ball. Silicon caulk is a useful adhesive for this. As the concrete is poured, simply screed off the top of the rail. The truck wheels will push the foam down and out of the way on the first charge. Both methods expose less track to corrosion and make the rail a lot easier to work and walk around, much less drive over.

Use standard 4-foot-long carts, unless the carts are made for specific lumber lengths or purposes. I prefer 6-inch side channels on the trucks rather than the standard 4 inches. These trucks last a lot longer, and the wheel is protected from forklift damage. Place bunk supports under each sticker, and you will find that a 4-inch square tube with end caps holds up longer for the bunks. Roller bearings are preferred, and make certain that the axles are equipped with grease fittings, accessible from the outside of the loaded cart.

If you are using carrier-style kiln doors, install lift-out rail sections at the thresholds, but make certain the foundations between the kilns and loading aprons are properly aligned and will remain so. Without these lift-out sections, expect track erosion under the doors. Examine the door carriers of your vendor and make certain you have plenty of leverage and support. All pivot and rolling points should have grease fittings for routine lubrication.

Barn-style doors have evolved to the point where I am starting to like them. When considering your purchase, take a close look at the threshold area of the potential vendors and visit a kiln in operation if you can. The threshold is the hardest area to seal, and there should be a minimum of light visible from the inside of a closed door. If the kiln is running, note how much steam escapes around the door in the early morning when the vents are closed. On many older kilns with barn-style doors, this is a very dangerous place to be standing.

Baffles

Use hinged baffles for the floors, overhead, and ends constructed of, at minimum, 16-gauge steel. Aluminum baffles may be available, but make certain that they are well restrained to keep them from flying back from the load when the fans are started. The lift cables on the overhead baffles must lift straight up, not out, to reduce the effort required to raise them. I avoid concrete floor baffles, since they do not provide a good seal and actually contribute to uneven airflow distributions top to bottom in the kiln. I prefer to use chains rather than springs to hold the end baffles in place, since they do not stretch out and will not snap back in you face. Solid rod braces are generally good as well. Baffles should never impede the flow of air through the lumber. They must contact the lumber stacks at their edges (top, bottom, and ends) for maximum, uniform airflow.

Make certain your supplier baffles the space between the floor and the bottom of the center coils. They must also baffle between the overhead coil and the outside wall on both sides.

Plenum baffles are useful in double track kilns, but not mandatory if the kiln is properly multizoned top to bottom on the center coil. Do not waste your money putting them in single track kilns.

Kiln Structure

Carbon steel structural members are fine, provided the kiln has a raised perimeter foundation. See that the primed steel of an erected structure is coated prior to installing the aluminum skin panels. If painted afterwards, there are large areas of the structure that will be missed by the painters and not seen by you until it is way too late for recourse. If an all-aluminum structure is available for your application, by all means consider it. All-aluminum is mandatory when drying oak and hemlock owing to the severe corrosion problems encountered with these species. Consider them also if your product mix demands long equalization and conditioning treatments.

Any portion of the kiln that is in a transition between the kiln interior and exterior must be either stainless steel or aluminum, since the condensation in these areas accelerates the corrosion of carbon steel. This applies particularly to all door frames, jambs, thresholds, and lintels. Exceptions are steam lines or fan shafts and supports for external fan drives.

Fan decks are preferably flat and capable of supporting the weight of a man without having to drag in planking to walk on. If you have internal kiln motors, they need to be able to support the weight of the motor and a man. If not, you will need to install an overhead rail to move motors to the kiln perimeter.

Man Access Doors

Ground-level access doors at each corner of the kiln are handy and convenient. They should open directly into the kiln plenum and not the end of the load, which would require you to unlatch the end baffles to gain access.

You will appreciate being able to enter the fan deck through an access door off an end wall catwalk or overhead control room. You should not have to squeeze through a trap door in the fan deck to do maintenance in this area. If you have internal kiln motors or a line shaft system, you will also need at least one door through the fan bulkhead to move between the two sides. During installation make certain that the motors of internal motor kilns are on the side of the bulkhead that the fan deck access door opens into. If you have purchased, against my recommendation, a through-the-roof external fan drive, the drive should also be on the side of the fan deck access door. Fan deck access doors to all other external fan drives should be on the opposite side of the drive with at least one door in the fan bulkhead. I personally prefer a bulkhead door between every other fan on these, so as to gain access to each drive for maintenance without having to crawl over, through, or around the drives.

Panel Design and Insulation

Panels must be designed to allow the escape of moisture, which enters as vapor and condenses inside. They must also prevent shifting or sagging of insulation material. Pay close attention to the panels on new kilns to detect bulges or hot spots on the outside. This is an indication that problems are beginning. Bring them to the attention of the manufacturer for resolution.

There is a continuing controversy over the merits of different insulation materials, panel thicknesses, and R-values. Let's talk about R-values first.

R-value is the resistance a material has to heat transfer. It is expressed by the equation:

$$\text{Heat loss (BTU/HR)} = \frac{(^\circ\text{F inside} - {}^\circ\text{F outside}) \times \text{Surface area in sq. ft.}}{\text{R-value}}$$

Let's look at an example comparing the heat loss from a double track by 68-foot high-temperature dry kiln with R-9 and R-20 insulation. A kiln of this size will have approximately 7,000 square feet of surface area and hold approximately 129 Mbf of lumber. Assume the kiln interior temperature is 240°F with an outside temperature of 65°F for a total temperature difference of 175°F.

For R-9:

$$\text{Heat loss} = \frac{175^\circ\text{F} \times 7{,}000 \text{ sq. ft.}}{9} = 136{,}111 \text{ BTU/hR}$$

For R-20:

$$\text{Heat loss} = \frac{175^{\circ}\text{F} \times 7{,}000 \text{ sq. ft.}}{20} = 61{,}250 \text{ BTU/hr}$$

Heat loss difference between R-9 and R-20 panels = 74,861 BTU/hr

or,

Total pounds of steam per hour @ 870 BTU/lb** = 86.0 lb/hr

** Latent heat @ 125 psig.

Average steam demand required to dry 129 Mbf of lumber in 20 hours:

$$\text{Average steam demand} = \frac{129{,}000 \text{ bf} \times 2.8 \text{ \# steam/bf}}{20} = 18{,}060 \text{ lb/hr}$$

$$\text{Heat loss through the panels as a percentage of average steam demand} = \frac{86.0 \text{ lb/hr} \times 100}{18{,}060 \text{ lb/hr}} = .48\%$$

In terms of the total heat consumption of this dry kiln, heat loss through the kiln skin is very low. If a mill is burning wood waste for fuel, it is insignificant. You can fill in your values for ambient temperature, capacity, drying time, etc. to arrive at an estimate for your situation. In very northern climates, heat losses through the kiln skin as a percentage of total average heat consumption might approach 5 to 6%, particularly with the longer drying species. The higher R-factor here becomes a very important consideration, particularly if the mill is burning fossil fuels to dry its lumber.

In more temperate climates, a more lasting concern is the long-term durability of the insulating material. After all, we do need insulation in our kilns. Rather than enter into a potentially litigious controversy over insulating materials, I will simply list what is available and give you some guidelines for selection. I have seen them all and have my own opinion. Let the salesmen slug it out. The following materials are available:

- fiberglass board
- foamed glass
- polyurethane foam
- polyisocynurate foam
- hybrid boards combining layers of different materials
- mineral wool

The insulation you select must be durable and not retain water. It must not lose its shape in the panel. It must be chemically inert to wood chemicals under high temperatures and saturated humidity conditions. It must not degrade over the life of the kiln. Ask to see kilns with the proposed panels that have been in operation for over a year. Inspect them personally. Look for bulges or distortions in the panel. Run your hand over the surface of the panel while the kiln is running to detect hot spots. Many salesmen will tell you that the foams do not absorb water. Perhaps not. However, water vapor passes freely through them and condenses when it reaches the dew point. Studies at the University of Tennessee[2] have documented moisture contents in these foams from 500 to 800%, which negatively impacts on their R-factor. This water is also a weak organic acid that corrodes aluminum. Look closely at the outside base of vertical panels for the white surface streaks that indicate corrosion from the inside. There are good panels out there, you just have to discover them.

One comment. Back when I did kiln surveys as an SPIB inspector, we often punched holes in the side walls of kilns to insert dial thermometers to check temperature balance. Never do this. Moisture-laden air pushes into the panels by the internal kiln pressure, and its moisture condenses on the interior of the outside panel. Water will soon saturate the insulation in that panel and reduce its effectiveness.

Steam Heating Systems

Mount control valves where they are easily accessible. I prefer to use flanged valves with Flexitallic gaskets for their ease of removal. All control valves must be fully modulating. While pneumatic control valves are the most common control valves, I do like electronic valves in certain situations. If you live in areas of high humidity or extreme cold, you may find these preferable to fighting air line freeze-ups and leakage. There is no need to install bypass steam lines around the control valves. In the old days these valves were useful. Today they are simply another item to leak and cause problems. A single shutoff valve ahead of the valve is needed if you have not installed a single large valve on the steam header. I prefer a high-quality ball valve for the smaller shutoffs rather than a globe or gate.

In terms of durability and economical heat transfer, it is hard to beat standard 2-inch schedule 40 finpipe. Always remember, no one ever complained about having too much heat in a dry kiln. If you are considering alternative finpipe, be very cautious. Being relatively expensive, aluminum and stainless steel finpipe is often overrated by the manufacturer in regards to heat transfer. This allows him to justify (in his mind) installing less linear footage of pipe. In every instance where I have looked at this, the mills have ended up with kilns with inadequate heating systems. Unless you are drying one of the more corrosive wood species, I recommend sticking with the tried and true.

Be leery of heating systems with coil headers smaller than 4-inch schedule 40 pipe. Smaller headers just do not seem to distribute the heat adequately to the individual coils, particularly when the 2-inch finpipe is involved.

Center heating coils on double track kilns should match the load area top to bottom and end to end. Gaps between adjacent coils or zones should not exceed more than 18 to 24 inches down the kiln length. There should be no spaces beyond normal coil spacing top to bottom. Coils must be run horizontally.

The only trap for dry kiln applications is the inverted bucket, primarily because it is the only trap capable of removing noncondensables (carbon dioxide and oxygen are the main offenders) from the steam. Sorry, there is no need to argue this point with me. There should be isolation valves and unions on the inlets and outlets of each trap, and the plumbing should be standardized to facilitate changing the trap. Strainers ahead of the trap are okay, although you can do just as well with a dropout and blow-down valve ahead of the trap inlet. If you do elect to install strainers, make certain they are equipped with an adequately sized blow-down valve. The trap should also be equipped with a valve on its test port, so the trap can be tested periodically. As with the control valves, I prefer a high-quality ball valve whenever a valve is indicated. I find check valves a nuisance and remove them whenever I encounter them. Oh, and by the way, don't forget to assemble your unions with Neversieze. Not only can you tighten unions tighter to seal leaks, but they are easier to break apart when the time comes.

The traps must be accessible. Kiln manufacturers who place their traps 10 to 20 feet off the floor should be soundly flogged. There are ways to route the drain lines and avoid condensate pooling problems without resorting to this. How can anyone routinely monitor their traps when they are installed in this manner? I continue to generally prefer traps inside the kiln, although I have seen some recent installations where they were mounted outside and worked well. Come to think of it, while we are flogging the manufacturers who place their traps in the air, we might as well round up the ones who insist on putting them in a pit. They should be shot.

Schedule 40 pipe is fine for all steam lines and drains up to the traps. Beyond the traps, consider using schedule 80 if the lines are laid in any ditch or nasty old pit. By the way, while the pipe fitter is routing lines in the fan deck area, have him put them out of the way so you do not have to crawl over, through, or under them simply to walk down the fan deck.

Steam mains must slope away from the boiler and be equipped with a drip leg and trap at the far end. Extremely long steam main runs will have multiple drip legs along their length. Steam headers must likewise have a drip leg and trap, and I like to slip this trap just inside the kiln wall to protect it in freezing weather. Steam mains and headers must be insulated and jacketed along their length. Make this a part of the kiln manufacturer's contract. Condensate return lines need not be insulated unless there are sections where personnel may come into contact with them.

Fan Systems

If your drying process calls for temperatures in excess of 200°F, then place your fan motors outside the kiln. Avoid the through-the-roof external drives in favor of those mounted at the eave. If your fan drive has more than two bearings, then insist on Cooper-style split bearings for all of those that are internal. The external bearing or the one closest to the sheave must be a thrust bearing. All others are floating. Never install external greasing systems on your internal fan motors or bearings. For a discussion on fan styles and required horsepower refer to the previous chapter on fan systems.

Humidity Systems

I have never encountered a power vent arrangement that satisfied me regarding its control accuracy. Stick to the roof-mounted vents. Whatever the vent style you end up with, its control must be fully modulated over its entire control range. When comparing vent systems between the various vendors, it is useful to calculate the total vent surface area each has to offer. You may be surprised at the variation you find. If the vents are fully modulated, you should have no problem with what may seem to be a tremendous amount of vent area. As with the heat system, no one ever complains about having too much vent capacity.

Sprays must likewise be fully modulated and adequately sized to the application. If conditioning and equalizing is a regular part of your drying operation and you are operating at pressures much in excess of 20 psig, then you should install a desuperheater to bring your high-pressure steam down to the lower pressure. This is best accomplished with a separate desuperheated manifold rather than the simple water injection systems commonly offered. If you are serious about conditioning and equalizing, you need the manifold system.

Controls

Rather than go into redundant details, please read the chapter on controls earlier in this book to obtain a sense of the features available in the marketplace. Above all, it is recommended that you obtain a user list from each potential vendor and call several of those listed to obtain a sense of what they have encountered since the installation of their unit. It is more helpful for you to speak to the person who actually runs the equipment than the individual in the office who might have made the decision to purchase it. Even so, keep in mind that the person you speak to likely has very little experience beyond the controls they are currently running, or the equipment they are currently running replaced controls that were so bad or outdated that anything would be an improvement.

To sort through the claims for yourself, visit several mills representative of each vendor and look at the lumber being produced. Get a feel for how well they are running their kilns. Ask if the mill experienced any tangible improvement with the installation of their controls. Did they experience reductions in steam usage or electrical energy? Did grade recovery improve? Did trim losses go down? Ask if they have any information on moisture distributions both before and after the controls were installed.

Ask the operator how long it took the vendor to get the kiln up and running once the startup technician was on site. Does he believe the vendor supplied sufficient training for him to effectively operate the control system? Did he find out later that certain items were unexpected "extras"? Ask to see and thumb through the operation manuals (documentation) for the system. Does it appear to be complete? Ask the operator how he feels about the documentation. How easy is it for him to speak to someone at the vendor's plant to get his questions answered? How well can the operator answer your questions?

Talk to the plant electrician or controls technician. Does he feel comfortable troubleshooting the system? What kind of problems has he had? Ask about the availability of software upgrades. Are they still running the control software they were started up on? Have they tried to get an upgrade? What happened if they did?

Finally, sit down at the computer and ask the operator to take you through the screens. How easy is it to retrieve charts and other information? How easy is it to obtain information from charges run one week ago? Last month? How about last year? Is the information useful? Have him walk you through starting up a kiln charge. Can you make keystroke mistakes that shut down the system? What sort of problems has he encountered?

Take your time selecting your new dry kiln. It is a major purchase, which you will have to live with for many years to come. Shortcuts now may haunt you for the next 20 years.

REFERENCES

1. Townsend, Kent. "Fire in the Dry Kiln." *FPRS Wood Drying News Digest.* Madison, WI: FPRS, 1981.
2. Little, R. L., and W. Moschler. "Dry Kiln Corrosion Research at the University of Tennessee." *Lumber Drying Sourcebook.* Madison, WI: FPS, 1998. pp. 311–318.
3. Armstrong. *Steam Conservation Guidelines for Condensate Drainage.* Three Rivers, MI: Armstrong International, Inc. Handbook N-101, 1991.

Preventive Maintenance for Steam Kilns

Daily Routine

	When you arrive for your shift try to find out if there were any problems affecting the dry kilns during your absence.
	Examine the charts for each running kiln to see that the schedules are running in control. Investigate any problems.
	Visually check water flow to the wet bulb trough. Walk around the kilns to see if anything seems out of place.
	Listen for steam leaks as you open a kiln and prepare to pull it. Locate each leak and schedule its repair according to severity.
	Check fit and action of the baffles as you walk down the kiln plenum. Schedule the repair of any damage you cannot easily repair.
	As you walk down the plenums reach up and feel the wet bulb sock. Replace it if it feels crusty or stiff. If the trough is overflowing, fix it.
	Remove any fallen boards from the floor and swing temperature sensors out of the way if necessary. Note any problems for repair.
	As the lumber carts are being pulled, note if any of the wheels are not turning and mark them for repair.
	Make a final inspection of the lumber going into the kiln. All lumber must have continuous support from top to bottom. Attach tracking tags.
	Make certain the baffles are secure for the new charge. Position the temperature sensors. Make certain the doors are closed properly.
	After startup of the next charge, verify that all the control valves have opened and are functioning. Vents should be closed
	Make certain all fans are running. Investigate the reason for those that trip out frequently. Listen for slipping fan belts.
	Spot-check sticker alignment and kiln setups. Spot-check lumber for manufacturing problems affecting your drying. Correct deficiencies.
	Lubricate kiln truck wheels on the carts being loaded for the next charge. Remove damaged carts for repair.
	Spend time at the planing mill spot-checking moisture content and looking for drying defects. Talk to the planer supervisor and graders.
	Drain water from air filters and lines.

Weekly Routine

	Replace your wet bulb socks unless you do it on a more regular basis.
	Check traps and control valves by opening the valves and manually testing each trap. Blow down mud legs or strainers ahead of the traps.
	Sweep out the kiln to remove accumulated dust.
	Do scheduled repairs from previous week—steam leaks, baffles, and so on.
	Record fan motor amperage for each motor on the system in your motor log.
	Grease kiln truck wheels unless you are doing it daily.

Monthly Routine

	Grease internal fan motors and bearings. Inspect fans and drive supports for damage. Don't forget to close all access doors when done.
	Tension and replace fan belts as required.
	Replace trap kits in the traps scheduled under your PM program. Inspect kiln walls and doors for deterioration and schedule repairs.
	Clean out the underground condensate line trench.
	Inspect vents and linkages, testing them through their full range of operation. Grease vent shafts on vents with internal linkages.
	Top off oil cups on line shaft kilns with cup oilers. Inspect the oiling system on any line shaft system.

Every Two Months

	Check external motor bearings and grease as necessary.

Every Three Months

	Lubricate internal vent line shafts.

Every Six Months

	Lubricate bearings and clevis pins on external vent actuators.
	On line shaft fan systems with cup oilers, replace the oiler wicks.

Once a Year

	Replace the trap kits on all traps on the main steam line and header drip legs.
	Inspect kiln roofs, walls, and doors for deterioration. Never coat the walls of an aluminum kiln.
	Inspect gaskets of all main loading and access doors. Replace where worn or damaged. Inspect door thresholds for damage.
	Remake all thermocouple tips.
	Inspect water supply lines to wet bulbs for signs of thinning. Replace if necessary.

Every Three Years

	Replace fan motor bearings.

Preventive Maintenance for Direct-Fired Kilns

Daily Routine

	When you arrive for your shift, try to find out if there were any problems affecting the dry kilns during your absence.
	Examine the charts for each running kiln to see that the schedules are running in control. Investigate any problems.
	Visually check water flow to the wet bulb trough. Walk around the kilns to see if anything seems out of place.
	Check fit and action of the baffles as you walk down the kiln plenum. Schedule the repair of any damage you cannot easily repair.
	As you walk down the plenums, reach up and feel the wet bulb sock. Replace it if it feels crusty or stiff. If the trough is overflowing, fix it.
	Remove any fallen boards from the floor and swing temperature sensors out of the way if necessary. Note any problems.
	As the lumber carts are being pulled, note if any of the wheels are not turning and mark them for repair.
	Make a final inspection of the lumber going into the kiln. All lumber must have continuous support from top to bottom. Attach tracking tags.
	Make certain the baffles are secure for the new charge. Position the temperature sensors. Make certain the doors are closed properly.

	After startup of the next charge, make certain that vents are closed.
	Make certain all fans are running. Investigate the reason for those that trip out frequently. Listen for slipping fan belts.
	Spot-check sticker alignment and kiln setups. Spot-check lumber for manufacturing problems affecting your drying. Correct deficiencies.
	Lubricate kiln truck wheels on the carts being loaded for the next charge. Remove damaged carts for repair.
	Spend time at the planing mill spot-checking moisture content and looking for drying defects. Talk to the planer supervisor and graders.
	Drain water from air filters and lines.
	Follow manufacturers' recommendations for burner and fuel handling equipment maintenance.

Weekly Routine

	Replace your wet bulb socks unless you do it on a more regular basis.
	Sweep out the kiln to remove accumulated dust.
	Wash accumulated ash from downcomers and the kiln interior in wood waste systems.
	Do scheduled repairs—burner, baffles, and so on.
	Record fan motor amperage for each motor on the system in your motor log.
	Grease kiln truck wheels unless you are doing it daily.
	Follow manufacturers' recommendations for burner and fuel handling equipment maintenance.

Monthly Routine

	Grease internal fan motors and bearings. Inspect fans and drive supports for damage. Don't forget to close all access doors when done.
	Tension and replace fan belts as required.
	Check distribution and supply ducts for missing bolts and damage. Turn on the recirculation blower and check again.
	Scrape any ash accumulation from the fan blades, being careful not to unbalance the blades.

	Inspect vents and linkages, testing them through their full range of operation. Grease vent shafts on vents with internal linkages.
	Follow manufacturers' recommendations for burner and fuel handling equipment maintenance.

Every Two Months

	Check external motor bearings and grease as necessary.
	Follow manufacturers' recommendations for burner and fuel handling equipment maintenance.

Every Six Months

	Lubricate bearings and clevis pins on external vent actuators.
	Inspect hammermill for wear.
	Follow manufacturers' recommendations for burner and fuel handling equipment maintenance.

Once a Year

	Inspect kiln roofs, walls, and doors for deterioration. Never coat the walls of an aluminum kiln.
	Inspect gaskets of all main loading and access doors. Replace where worn or damaged. Inspect door thresholds for damage.
	Remake all thermocouple tips.
	Inspect water supply lines to wet bulbs for signs of thinning. Replace if necessary.
	Follow manufacturers' recommendations for burner and fuel handling equipment maintenance.

Every Three Years

	Replace fan motor bearings.

FAN OUTAGE LOG

KILN #	FAN #	LOSS DATE	REWIND OR NEW	SERIAL NUMBER	HOW OLD?	CAUSE

FAN AMPERAGE LOG—KILN # ______

MOTOR HORSEPOWER: ______ FULL LOAD AMPS: ______

	DATE:		DATE:		DATE:		DATE:	
FAN #	KILN °F	AMPS	KILN °F	AMPS	KILN °F	AMPS	KILN °F	AMPS

Note: Gradually increasing amperage may indicate impending bearing failure. Unusually low amperage may indicate belt loss on belt-driven systems. Bouncing amperage indicates loose belts or excessive sheave wear.

TRAP SERVICE LOG—KILN # ______

TRAP #	TRAP SIZE	ORIFICE SIZE	SERVICE DATE	SERVICE DATE	SERVICE DATE	SERVICE DATE

Note: How frequently you rebuild your traps depends on how hard you run your equipment. On new kilns, pull down a selected trap (bottom center coil is a good choice) and examine orifice for wear. If the hemispheric ball that closes the trap begins to resemble a nipple, it is way past time for rebuilding. Once you determine how long it takes to get to this wear stage, simply begin rebuilding one or two traps a month until you can begin the process all over again. This spreads out the work and helps avoid losing traps unexpectedly.

Glossary

ACCEPTANCE SAMPLING A statistical quality control method that seeks to determine the quality of a product by sampling a small portion of it. It does not ensure that 100% of the product is acceptable, but it reduces the likelihood of acceptance of a defective product (high moisture content).

ADHESION The property of a substance (water) where it tends to adhere to other substances.

AIRFLOW The pattern of air movement within the dry kiln. This pattern varies with kiln design, loading methods, and level of maintenance and operation.

AIR VELOCITY The speed of airflow through the stickers, crossouts, and around the lumber expressed in feet per minute (fpm) or meters per second (mps).

ANISOTROPIC The characteristic of a material (such as wood) to have different properties depending on the direction being tested.

ANNUAL RING One year's growth increment of a tree composed of one band of springwood (earlywood) and one band of summerwood (latewood). Also called a growth ring.

AVERAGE AIR HEAT CONTROL Controlling the temperature in a kiln based on the average of both the entering and exiting air. Commonly used in kilns with long air passes such as package kilns.

AVERAGE MOISTURE CONTENT Given a group of moisture content values the average is the sum of that group divided by the quantity of moisture content values in the group. Also called the mean.

BAFFLE A metal, concrete, or wooden barrier used to deflect and control airflow through the lumber or around heating coils.

BAFFLE, CONCRETE FLOOR A raised concrete curb or walkway that serves as a stationary floor baffle.

BAFFLE, END A hinged baffle that closes the gap at the end of the lumber next to the doors.

BAFFLE, FLOOR A hinged or stationary baffle that closes the gap at the bottom of the lumber unit, preventing air from escaping under the load.

BAFFLE, OVERHEAD A hinged baffle that closes the gap between the overhead fan deck and the top of the lumber load.

BAFFLE, PLENUM A hinged or stationary baffle that hangs vertically in the plenum chamber and serves to split and control the airflow top to bottom within the dry kiln.

BARK The outside layer of a tree stem. It provides protection, insulation, and reduces moisture loss.

BASELINE MOISTURE PROFILE Data gathered from a dry kiln and used to compare how the moisture content variation relates to the air velocity variation top to bottom. Used to identify mechanical drying problems within a kiln.

BOILER Steam generation equipment.

BOLSTER *See* CROSSOUT.

BOUND WATER Water that is bound within the cell wall of wood. Water held in wood below the fiber saturation point. Bound water is removed by diffusion.

BOW A deviation flatwise from a straight line drawn end to end on a piece of lumber. *See also* WARP.

BROWN STAIN A brown to chocolate brown stain found in several softwood species and caused by the concentration and modification of extractives.

BULBS General term referring to the dry bulb and wet bulb sensors within a kiln.

BUNK *See* CROSSOUT.

BUNK SUPPORT The horizontal kiln truck supports on which the lumber is placed for drying.

CALIBRATION Adjusting the control or recording equipment to reflect the actual conditions in the kiln. Commonly done for temperature sensors, I/P transducers, and P/I transducers.

CAMBIUM The layer of tissue dividing the bark from the wood, which forms new bark to the outside and new wood to the inside as the tree grows. This is the actual living portion of the tree stem.

CAPILLARY ACTION or FLOW The combination of solid-liquid adhesion and surface tension by which liquid moves through a cellular structure.

CASE-HARDENING A condition of stress and set in dry wood in which the outer fibers are under compressive stress and the inner fibers under tensile stress. The stresses persist after the lumber is dry and cause warp if the wood is remachined after drying. *See also* REVERSE CASE-HARDENING.

CHECKS A splitting of the wood fibers within or on a log or lumber. Checks result from uneven shrinkage. They occur on the surface, internally, and on the ends of lumber.

COHESION The property of a substance (water) where it tends to adhere to itself.

COILS, STEAM The radiant heating surface within steam-heated kilns. Steam coils contain the steam and transfer its heat through the coil wall to heat the kiln air, producing condensate in the process.

COILS, CENTER The steam coils that are positioned between the kiln tracks. Also called booster coils.

COILS, OVERHEAD The steam coils positioned overhead at or on the fan deck.

COLLAPSE Flattening or buckling of the wood cells during drying, resulting in excessive shrinking and/or uneven shrinking plus a corrugated surface.

COMPRESSION WOOD Reaction wood in coniferous (softwood) trees, which forms on the underside of leaning trees and limbs. This wood is characterized by high longitudinal shrinkage as compared to normal wood.

COMPUTER SIMULATIONS Computer software that models actions or occurrences in the real world. By using modeling or simulation techniques on the computer, you can experiment with various drying techniques without the risk of damaging lumber.

CONDENSATE Water formed by removing latent heat from steam within the kiln heating coils.

CONDITIONING A process for relieving the stresses present in wood at the end of drying.

CONTROL VALVE A device that controls the flow of steam to the steam coils.

CORE, LUMBER The central portion of a board cross section. The interior of a board.

CROOK A deviation edgewise from a straight line drawn end to end on a piece of lumber. *See also* WARP.

CROSSOUT Horizontal spacer between units of lumber. These align over the stickers and serve to separate the units for forklift handling. Also called bolsters and bunks.

CUP Deviation in the face of a piece of lumber from a straight line drawn edge to edge. *See also* WARP.

DEAD PILE To stack lumber in units without stickers.

DEGRADE Drop in lumber grade due to manufacturing or drying practices.

DENSITY The weight of wood in relation to its volume. If not measured directly, density is inferred from the number of rings per inch. The more rings per inch, the higher the density.

DESUPERHEATER A device for removing the excess heat in steam as its pressure is reduced.

DIFFUSION Movement of moisture from areas of high moisture content to areas of low moisture content.

DIMENSION LUMBER Lumber manufactured for building construction.

DOOR CARRIER A device that lifts kiln doors and moves them away from the kiln opening along an overhead track.

DOORS, ACCESS Small doors, usually located at ground level but also found opening into the fan deck, which afford the operator access to and escape from the kiln interior.

DOORS, KILN Large openings through which lumber passes when loading a kiln.

DOWNCOMER Small, vertically oriented distribution ducts within direct-fired kilns that distribute heated air between the kiln tracks from the overhead main distribution duct.

DRESSED LUMBER Any dry lumber product that has been processed through the planer to finished dimensions.

DRY BULB A sensing device (thermometer) that measures the temperature of the air and is used to control drying conditions within the kiln or kiln zone.

DRYING RATE The speed at which water is given up by the drying lumber. Usually expressed as a percentage of moisture loss per day or hour.

EARLYWOOD Thin-walled wood cells deposited by the cambium during the fast growth of spring. Also called *springwood.*

EDGING Trimming the width of a board to remove defects.

ENTERING AIR CONTROL Sensing and controlling the heat output of a coil or heating zone by measuring the temperature of heated kiln air as it enters the lumber load.

EQUALIZATION The process of narrowing the moisture distribution between boards at the end of a drying charge.

EQUILIBRIUM MOISTURE CONTENT (EMC) The moisture content wood attains naturally at a given dry bulb and wet bulb depression.

EVAPORATE To change water from a liquid to a vapor form.

EXITING AIR CONTROL Sensing and controlling the heat output of a coil or heating zone by measuring the temperature of the heated air as it leaves the lumber load.

FAN DECK The overhead area where the fans are mounted.

FAN PITCH The angle of the fan blades generally measured one-third of the fan radius in from the tip.

FAN SYSTEM The air-moving equipment within a dry kiln.

FANS, CROSS SHAFT Fans with their shafts oriented perpendicular to the length of the kiln. Motor drives can either be through-the-roof or eave mounted.

FANS, DIRECT-DRIVE Fans mounted directly on the fan motor shaft.

FANS, LINE SHAFT Multiple fans mounted on a single shaft, which extends the length of the kiln. Used in conventional kilns.

FANS, SINGLE-SPEED Fans with motors that turn at one speed.

FANS, TWO-SPEED Fans with motors built with two sets of windings and capable of two rotation speeds.

FANS, VARIABLE-SPEED Fans whose motors turn at infinitely variable speeds by varying the line frequency of the power source.

FIBER SATURATION POINT (FSP) The moisture content at which moisture is saturated within the cell walls of wood and the cell cavities are free of water. For most species, this averages around 30% MC. Below FSP, water is held in wood as bound water within the cell walls. Above FSP, water is held as bound water and as free water within the cell cavities or lumen.

FINISH LUMBER A collective term for upper grades of lumber (generally boards) suitable for natural and stained finishes.

FINPIPE See COILS, STEAM.

FLITCHES A portion of a log sawn on two or more sides and intended for remanufacture.

FORKLIFT UNITS Lumber bundled or stacked into packages that can be lifted by a forklift.

FREE WATER Water held within the cell cavities of wood. Its movement is primarily by capillary action.

GRAIN The fibers in wood and their direction, size, arrangement, appearance, or quality.

GREEN LUMBER Lumber fresh from the saw.

HEARTWOOD Wood extending from the pith to the sapwood of a log or board. It is typically infiltrated with extractives, resins, and gums, which often make it darker in appearance than the sapwood. Heartwood is very pronounced in many species. Generally lower in moisture content than the outlying sapwood, it is less permeable to moisture movement.

HONEYCOMB Checks, not always but occasionally visible from the surface, that occur in the interior of a piece of wood, usually along the wood rays.

HYDRAULIC To move or convey by fluid.

IMPLOSION The sudden collapse of kiln wall panels due to unequal air pressure between the inside and outside of the kiln. This usually occurs on startup as hot, expanded air cools and contracts as it passes through green lumber. The vents must open on startup to prevent this.

ISOTROPIC The characteristic of a material (such as steel, aluminum, or glass) whose properties are the same no matter which direction it is tested.

JUVENILE WOOD Wood deposited next to the pith and characterized by high longitudinal shrinkage when compared to wood laid down as the tree matures.

KILN CARTS A framework fabricated of at least two kiln trucks and at least two bunk supports on which lumber is loaded for drying within the kiln. Kiln carts are required for track loaded kilns.

KILN KINK The distortion in lumber that results from the uneven forces applied by misplaced truck supports and crossouts. Once dried into the lumber, the effect is permanent.

KILN TRUCKS A two-wheel assembly, two of which are required to build one kiln cart.

KILNS, ACCELERATED Kilns that operate at temperatures above conventional temperatures (usually 200°F) but below high temperatures (above 212°F).

KILNS, CONVENTIONAL Kilns that operate at temperatures below 200°F

KILNS, DIRECT-FIRED Kilns that are heated by blowing the products of combustion into the kiln.

KILNS, HIGH-TEMPERATURE Kilns that operate at temperatures above 212°F.

KILNS, PACKAGE A trackless compartment kiln for drying packages of stickered lumber. It has large doors through which a forklift loads lumber directly within the kiln. Also called a *side-loaded kiln* or *sideloader.*

KILNS, PROGRESSIVE A relatively long track kiln in which green lumber is introduced at one end and dry lumber removed at the opposite end. These kilns are characterized by a very constant and stable steam flow. An old concept, these kilns are rarely, if ever, encountered in the United States owing to the problems in maintenance, startup, quality, and cost.

KILNS, STEAM Kilns heated by steam.

KILNS, THERMAL OIL Kilns heated by thermal oil.

KILNS, TRACK Kilns loaded by placing lumber on kiln carts that run on tracks within the kiln. These can be single, double, and occasionally triple track kilns.

LATEWOOD Thick-walled cells deposited by the cambium layer following the fast growth spurt of spring. Also called *summerwood.*

LIGNIN The thin, brown cementing layer between adjacent wood cells. This is the material removed during pulping to separate the fibers and produce paper.

LONGITUDINAL DIRECTION Oriented in the direction of the length of the wood grain.

LUMEN The cavity of a cell.

MANUFACTURE Any of the various processes that convert logs into lumber. *See also* EDGING, TRIMMING.

MISMANUFACTURE Wood imperfections or blemishes that result from the manufacturing process. *See also* RABBETED EDGE, SAW OFFSET, SNAKED SAWS, TAPER.

MODULATE To control within an infinite range between 0% and 100% as opposed to on/off control. This is the preferred control method for all aspects of the drying process.

MOISTURE CONTENT For lumber, the weight of the water contained in wood, expressed as a percentage of the weight of the oven dry wood. Known as the oven dry basis. For boiler fuel, the weight of the water contained in wood, expressed as a percentage of the weight of the green wood. This is called green or wet basis moisture content.

$$\text{Moisture content}_{\text{oven dry basis}} = \frac{(\text{Green weight} - \text{Oven dry weight})}{\text{Oven dry weight}} \times 100$$

$$\text{Moisture content}_{\text{wet basis}} = \frac{(\text{Green weight} - \text{Oven dry weight})}{\text{Green weight}} \times 100$$

MOISTURE DISTRIBUTION. 1. Variation in moisture content within a board. 2. Variation in moisture content between boards.

MOISTURE GRADIENT The difference in moisture content between areas within a board. Usually, this refers to the moisture content difference between the surface and core of a board.

MULTIZONE Control technique of dividing a kiln into more than two zones of heat control. Used in steam and thermal oil kilns.

OVERDRY Lumber dried to the point of having low moisture content and incurring excessive degrade.

PARENCHYMA CELLS Tissue active in the storage and distribution of carbohydrates; cells mostly short, with numerous simple pits.

PITCH 1. Degree of angle on a fan blade. 2. Accumulation of resin in the wood. This causes resistance moisture meters to read high. 3. To hurl horseshoes at a distant stake embedded in a sand pit.

PITH The small, soft core occurring in the structural center of a tree trunk, branch, twig, or log.

PITS A recess in the secondary wall of a cell where a thin membrane may permit liquids to pass from one cell to another.

PLC Programmable Logic Controller/Process Logic Controller. Industrial computer with I/0 capability. Generally programmed in ladder relay logic.

PLENUM CHAMBER The space (walkway) between the lumber and the outside walls of the kiln. This is the area pressurized by the fan system to deliver air flow to the lumber stacks.

PNEUMATIC To move or convey an object with air pressure.

PSYCHOMETRIC CHART A table or graph employed to determine the relative humidity of air from observed wet and dry bulb temperatures.

QUALITY CONTROL Any activity which helps to maximize the value of the raw material and lumber products through all phases of the manufacturing process.[1]

RABBETED EDGE Saw offset on the edge of lumber, which is characteristic of Chip-n-Saws. Graded by minimum wane rules.

RADIAL A longitudinal surface or plane extending wholly or in part from the pith to the bark.

RAYS A ribbonlike aggregate of wood cells extending radially across the grain, radiating out from the pith to the bark.

REACTION WOOD Wood that forms in a tree in response to the tree leaning. Called compression wood in softwoods, since it forms on the underside of the stem or limb. In hardwoods it forms on the upper side of the stem or limb and is called tension wood.

RELATIVE HUMIDITY The technical definition is: The ratio of the partial pressure of the vapor in a mixture to the saturation pressure of the vapor at the same temperature and pressure of the mixture. For our purposes: The ratio of the weight of water vapor in a given unit of air compared with the weight the same unit of air is capable of containing when fully saturated at the same temperature. That is, air is capable of holding a certain maximum of moisture (100%) at any given temperature. How much it actually holds in relation to how much it could hold is the relative humidity. It is expressed as a percentage between 0% and 100%.

REVERSAL, FAN The act of changing fan directions to move airflow through the lumber in the direction opposite to the way it had been going. Fans are either running FORWARD or REVERSE. Which is which is purely arbitrary, and each kiln manufacturer has adopted their individual conventions. All temperature sensing and venting is affected by fan direction

REVERSE CASE-HARDENING A final stress-and-set condition in dry lumber in which the outer fibers are under tensile stress and the inner fibers are under compressive stress. This condition is not reversible. *See also* CASE-HARDENING.

RIP MILL A manufacturing center that edges and trims rough lumber after drying to remove defects before grading and sorting for the planer mill.

RISERS Vertical distribution ducts within a direct-fired kiln that distribute heated air between the tracks from an underground main distribution duct.

ROUGH LUMBER Lumber, green or dried, that shows saw or machine marks on its surface.

RTD Resistance Temperature Diode. A temperature measuring device whose resistance to an electrical current varies with temperature. The resistance is measured to determine temperature.

SAPWOOD The outer layers of a tree stem directly under the cambium layer. The width of the sapwood varies with species and is generally lighter than the heartwood that it encloses. It is characterized by being generally higher in moisture content than the heartwood and comparatively permeable to moisture movement.

SAW KERF The width of a saw cut as the saw passes through the log or board.

SCHEDULE The "recipe" for drying lumber that dictates wet bulb and dry bulb set points throughout the drying cycle.

SET POINT The temperature to which you set your control instruments within the drying schedule.

SHEAVES Pulleys mounted on the end of the drive and motor shaft, over which belts pass to transfer rotational force from the motor to the shaft.

SHELL, LUMBER The outer portion of the cross section of a board nearest the surface.

SHRINKAGE The contraction of wood to water loss below fiber saturation point. Expressed as a percentage of the green dimension.

SHROUD, FAN The curved perimeter of a fan assembly that serves to keep the air from slinging off the ends of the fan blades and helps direct airflow through the fan.

SIGNAL A control impulse from a control device or sensor.

SILLS Wooden, concrete, or steel supports placed on the ground or pavement onto which lumber is placed for drying or storage.

SNAKED SAWS Wavy saw line seen most often in band-sawed lumber. Graded as skipped dressing.

SOCK, WET BULB A fabric wick placed over a temperature sensing element (dry bulb) hung over a water trough. The saturated fabric cools the dry bulb by evaporation, making it a wet bulb. How much the evaporation cools relative to the dry bulb temperature in the kiln is used to determine the relative humidity of the kiln environment and control the vents and sprays.

SORTING Segregating rough-sawn lumber into lots for stacking, drying, and marketing purposes. Lumber is sorted for species, length, thickness, width, and occasionally moisture content. Dressed lumber is also sorted in the planing mill.

SPECIES A group of plants or animals of a particular kind. Most trees are sorted by species prior to processing through the sawmill. Each species typically has its own optimum drying schedule.

SPIRAL GRAIN Characteristic of some trees where the wood fibers spiral around the tree rather than running straight up and down. Causes slope of grain in lumber.

SPRINGWOOD *See* EARLYWOOD.

STACKER A mechanical device for making up unit packages of lumber for drying. As each layer of lumber is deposited on the unit, stickers are inserted manually or by machine between the layers.

STANDARD DEVIATION Root mean squared deviation of the observed numbers from their arithmetic average. A statistical technique used to describe the variation in a group of related numbers. For our purposes, 95% of the group (population) will fall within ± 2 standard deviations of the average of the group (population).

STEAM Vapor produced by heating water above its boiling point.

STEAM SPRAYS Mechanical equipment for introducing water vapor or steam to the kiln environment. Used to raise the relative humidity.

STICKER A wooden, plastic, or metal strip laid between each layer of lumber as it is stacked for drying. Permits transfer of heat into and removal of water from the lumber stacks.

STICK LAYER The person who lays stickers.

SUMMERWOOD *See* LATEWOOD.

TANGENTIAL A surface plane that is perpendicular to the radial plane in wood.

TAPER Boards that are thicker or wider on one end than on the other. Graded as skip.

TARGET SIZES The sawn dimension of lumber that the mill endeavors to attain. Target sizes are composed of the finished lumber dimension plus allowances for planing, shrinkage, and variation in sawing accuracy to achieve skip-free lumber at the planing mill.

TEMPERATURE DROP ACROSS THE LOAD (TDAL) The difference in temperature between the air as it enters a package of lumber and as it leaves. This value is a measure of the heat demand of the lumber at that point in the schedule and is an indirect measure of moisture content.

THERMAL OIL A heat transfer fluid used instead of steam. Thermal oil and steam are not interchangeable within the same kiln. Thermal oil is capable of higher temperatures at lower operating pressures than steam and is not susceptible to freezing.

THERMOCOUPLE A temperature sensing device composed of two dissimilar metals that produce an electromotive force (emf) that varies with temperature.

TIP CLEARANCE The distance between the end of a fan blade tip and the enclosing shroud.

TRACHEIDS, LONGITUDINAL The elongated cells or fibers that compose the majority of the anatomical structure of softwoods. Also present in some hardwoods.

TRACKS The steel rails on which kiln trucks rest and that serve to guide the lumber through the kiln as it is being loaded for drying.

TRAM Term generally used to refer to a kiln cart that extends the entire length of a track kiln.

TRANSDUCER A device that converts power or a signal from one system to another. Typically, I/P transducers are used to convert electrical computer signals to pressure signals to move valves and actuators. P/I transducers are typically used to convert pressure signals in the field to electrical signals the computer can understand.

TRANSFER CARS A mechanical car for moving stacked units of lumber from one rail or track section to another. Used in transfer handling systems.

TRAP KIT The internal parts of a trap that are replaced during servicing.

TRAP ORIFICE The internal opening within a trap through which condensate is discharged.

TRAPS A device that separates condensate and air from the steam within the steam heating coils.

TRAPS, FLOAT AND THERMOSTATIC Traps that discharge condensate at a rate dependent on the position of an internal float. Air is removed through a thermostatic air vent until steam reaches the trap.

TRAPS, INVERTED BUCKET Mechanical traps that operate on the difference in the density between steam and water.

TRAPS, ORIFICE Traps that discharge steam at a constant rate and have no internal moving parts to alter or stem their flow. Simply a hole in a plate mounted in the condensate drain line.

TRAPS, THERMOSTATIC Traps that use an internal bellows, which expands when steam flows to the trap, closing the orifice. As condensate accumulates, the bellows contracts to release it.

TRIMMING The manufacturing step that squares the length of boards.

TRUCKS *See* KILN CARTS.

TWIST A deviation, flatwise, or a combination of flatwise and edgewise, in the form of a curl or spiral. *See* WARP.

VENTING The act of discharging hot, moisture-laden air from the kiln.

VENTING, CONVENTIONAL Venting by opening both rows of vents together using a single control output.

VENTING, DEMAND Venting method by which the exhaust vents modulate open to maintain wet bulb set points. Failing this, the intake vents begin modulating open. Used in kilns with computerized controls.

VENTING, EXHAUST Venting by opening the exhaust vents only. Used in conventionally controlled high-temperature kilns,

VENTS The opening(s) in the kiln roof or wall that can be opened and closed to control the wet bulb temperature in the kiln.

VENTS, EXHAUST Vents located on the pressure side of the fans.

VENTS, INTAKE Vents located on the low-pressure side of the fans.

VESSELS Tubelike structures of indeterminate length in porous hardwoods.

WANE The common usage refers to bark edges on boards, but it also refers to the absence of wood on the edge of a board from any source.

WARP Any deviation from a true or plane surface in a board. *See also* BOW, CROOK, CUP, TWIST.

WATER BOX or TROUGH A water container mounted under the wet bulb, which supplies water to the wick or sock.

WATER HAMMER Potentially damaging condition in steam pipes wherein waves develop on the condensate surface within a pipe as steam rushes over it. If the peak of the wave becomes high enough to reach the roof of the pipe, the water in the wave is pushed violently to the far end of the pipe by the steam flow.

WATERLOGGED Condition of steam coils wherein condensate is not discharged at a sufficient rate to keep it from accumulating in the coils. Filled with condensate, the coils fail to transfer heat to the kiln.

WET BULB A dry bulb with a wet sock pulled over it to measure the cooling effect of moisture evaporation from its surface. Used to monitor and maintain humidity conditions within the kiln.

WET BULB DEPRESSION The difference in wet bulb and dry bulb temperatures. When related to the dry bulb temperature, the relative humidity and EMC values within the kiln can be estimated.

ZONING The way a kiln is divided into temperature control areas.

REFERENCES

1. Brown, T. D. *Quality Control in Lumber Manufacture.* San Francisco: Miller Freeman Publications, 1982.
2. Forest Products Research Society. *Glossary of Terms Related to the Drying of Wood.* Madison, WI: FPRS, 1977.
3. Grant, E. L. *Statistical Quality Control.* New York: McGraw-Hill, 1964.
4. O'Dell, L. R. "Steam Traps—Basic Facts and Energy Saving Tips." *Heating/Piping/Air Conditioning* No. 5 (May 1977): 49
5. Rasmussen, E. F. *Dry Kiln Operator's Manual. Agricultural Handbook No. 188.* Forest Service, GPO, 1961.
6. Southern Pine Inspection Bureau. *Grading Rules.* 1977 Edition. Pensacola, FL: SPIB, 1977.

Index

Boldface numbers refer to pages with figures only on the relevant subject.